I0489854

A Welder's Handbook to Robotic Programming

Published by Timothy Craig Everhart

Copyright 2014 Timothy Craig Everhart

Cover Design by Laura Shinn Designs

ISBN-13: 978-1499396898

ISBN-10: 1499396899

This book is licensed for your personal enjoyment only. This book may not be re-sold or given away to other people. If you would like to share this book with another person, please purchase an additional copy for each recipient. If you're reading this book and did not purchase it, or it was not purchased for your use only, then please return to Amazon.com and purchase your own copy. Thank you for respecting the hard work of this author.

Table of Contents

Always work with a Copy
Shifting programs to a different grid location
Shifting part of a program
A single Change at a time
Call Programs
Move programs to another station
Temporary Continuous mode
Using the Shift function to write programs Quicker and Easier

New Program
Job Begin
Job End
Check Weld
Troubleshoot weld out of place
End of Day

Dedications

I give my sincere thanks to my good friends Richard Gordon, George Holt and the rest of the staff at Metal Works. Were it not for their great need I may have never written this much needed book.

Another person who played a very significant role in this endeavor was Robert Mann, the "Every Mann". It was some of the things that he said that led me down the path to becoming a robot programmer in the first place.

A special thanks also to everyone at Capitol Robotics where most of the robots that I have programmed and the instruction that I have received came from.

I also have to thank my beautiful wife Michelle and my three great kids, Angela, Chelae and Anthony. Without their love and support I would never get any writing done.

Preface

This book has been a hard book to write because I have been compiling it from two completely different perspectives at once. I've done this because I wanted this book to be useful to the owner, buyer, or potential buyer of a welding robot as well to the programmers of the welding robots. Both of these groups have a need for a book that doesn't seem to exist and it just seemed to make more sense for me to write one book instead of two.

The first part of the book is geared more towards the owner/operators of weld shops that have a need of a robot and the rest is directed more towards the workers that are trying to learn how to program the robots.

My hope is that both groups will find a use for the entire book but I'll admit that there is going to be much more for the people that need to learn how to program the robots. Their need is much greater.

You will find that some things are mentioned several times and in multiple chapters of this book. Do not feel that I am insulting your intelligence when I do this, nothing could be further from the truth. When you see something repeated in this manner it is because that I feel that its importance is such that it bears repetition. The reason that I do this is because the more times that you see, read, hear or do something the more likely it is that you will actually remember or even learn it. Repetition has always been utilized as a valuable teaching tool in any type of education from preschool to the advanced university course studies.

I also hope that some of the curious souls that end up with this book in their hands for whatever reason, especially any of the younger generation of welders, that this book may get them interested enough for them to give serious consideration into becoming a robotic programmer.

I'm also looking at the real possibility of this book being used in many of the training programs that are teaching robotic programming because there is no text book out there that they can use, at least until now.

But whoever you are or for whatever reason that you have picked this book up, I hope it will give you a better understanding about robot programming and the growing need for robots and their programmers in the modern work place.

Introduction

Can just anyone learn to program a robot? Absolutely they can. It doesn't take a rocket scientist only a decent amount of math sense and maybe a touch of computer experience along with a good bit of stubbornness. Can just anyone become a top notch programmer? They most certainly cannot. The device that you use to program a robotic welder is called a teach pendant for a very good reason, you have to teach the robot how to weld each and every joint. This means that to get a robot to produce top quality welds you need to first know how to produce those top quality welds yourself.

When I'm welding by hand I have often had people look at my welds and remark that I have been programming robots for so long that my welds are beginning to look like they were done by a robot. That statement is completely, one hundred percent backwards. They don't seem to grasp the concept that if I didn't have the ability to make welds like that there would be no way that I could possibly tell the robot how to do it. So although nearly anyone with a moderate amount of intelligence can actually learn how to do the basic programming, becoming a top notch programmer does come with one prerequisite, you need to first be at least a good welder.

There's also another factor that you have to take into consideration, when a programmer is tweaking the welds in a robot program they need to be able to examine a weld and diagnose the problem with the bead as well as knowing what should be adjusted to correct the problem. Troubleshooting a weld is next to impossible for anyone except an experienced welder; you can only do so much with a troubleshooting guide and some pictures.

Another fairly common misconception is that anyone that can operate a robot welder can learn to be a good programmer as well, I'm afraid that this is just not the case. One of the big selling points for robotic welding systems is the fact that once the programming is in place just about anyone can be shown how to operate a robot. All you have to do is be able to remove the finished product from the fixture, load the next round of parts into the fixture and then push the green button again. The only prerequisite to being able to learn to operate a robot welder is a small dose of common sense. If the robot stops or fails to start when the green button is pushed then the operator calls for help and the programmer comes to get it going again. It's the same deal if a robot stops unexpectedly in the middle of a run, the operator can be taught how to pull the torch back and do a preliminary check of the tip and nozzle, even to change them, but if it turns out to be something else they can't fix it and will have to go get the programmer.

If you have ever watched a George Jetson cartoon and saw him come home at the end of the day with a sore finger from doing nothing except pushing a button, that day has finally arrived in the guise of a robot operator. You may think that what I am about to tell you is a joke; I agree that it does sound like one, but I have actually taught someone how to operate a robot that was still having problems operating the punch clock to get clocked in and out every day. A single programmer can easily take care of several robots, if it wasn't for the possibilities of needing the programmer on a day they were off for some reason, you wouldn't need a second programmer. Even with multiple robots, the times when two of them need to be programmed simultaneously would actually be a rather rare occurrence.

Each year robots are finding their way into even more industries. One manufacturing trade where robots have become deeply seated is welding. In the last couple of decades the auto industry has become saturated with welding robot arms, in fact, welding is one of the most widespread and successful applications of industrial robots. Although the auto makers have a few MIG and TIG robots the majority of them are still of the SPOT welding variety. MIG welding robot arms are now becoming even more popular than their spot welding cousins. They are finding their way into smaller and smaller shops; I've seen several cases where even a business with less than ten employees can now afford to purchase one of these technological wonders. MIG welding robots are becoming so affordable that with the right programming it's possible for one good sizable order to sometimes pay back the entire investment with interest. In fact, I know one such facility with less than forty employees that has three welding robots. Each one of their robots was paid for out of the profits of the original order that prompted the purchase. One could say that the company didn't even purchase the robots themselves; the customer actually purchased the robots for them. I personally can't think of a better way to buy a robot or any other piece of equipment for that matter.

Automated welding has been popular in the manufacturing industry for many years because of the cost savings they offered but the older 'Hard Setups' were only cost effective in huge runs of several thousands of units. The entire setup was only for one job and the time and cost of building it had to be compensated for by an extremely high number of units built and/or long periods of usefulness. At the end of the production period the entire setup, except for the welding equipment, was usually scraped and another built for the next equally large job.

Robotic welding, on the other hand, is just as flexible as the 'hard setups' are ridged. The smaller jobs that have only a few welds can be programmed so fast that the only cost that must be compensated for is the construction of a fixture. Also, because of the robot's ability to quickly tack weld before starting the actual welding, the fixtures can be much simpler and therefore have a lower cost than the fixtures used in the hard setups as well. This makes the welding robots a much better choice and it is the reason that they are beginning to saturate the niche industries that are flourishing today.

The number of welding robots that are being used in the manufacturing field today is already a staggering number and it is constantly growing. However, the number of manufacturing facilities that need to utilize robots to increase their productivity is even larger. If even a small fraction of these actually purchase a robot in the next few years there is going to be a great need for more programmers of these robots that are entering the workplace. In some areas of the country there is already a noticeable shortage of trained programmers.

Many small manufacturers that have looked at the possibility of buying a welding robot are asking themselves the wrong question. They are looking at it from the perspective of "Can I afford to purchase a robot?" They should instead be asking, "Can I afford not to buy one?"

In fact, the big question of the twenty-first century for all metal fabricators large and small alike is not IF they are going to adopt robotic welding technology but WHEN. Automation is the lifeblood of any manufacturing operation and robot welding systems are the wave of the future for metal fabricators, they are either going to have to catch the wave and ride it or be drowned by it.

The world we live in today is not the laid back, easy going world of our grandfather's day. We live in a fast paced, highly competitive society with an equally fast paced and competitive marketplace. If you are in the manufacturing industry then there is always someone who is trying to steal your customers. The keys to keeping those hard won customers are the speed of delivery, high quality products, and low overhead costs. Most changes that you can make towards improving a business will only affect one or possibly two of those keys but a welding robot can actually make a huge difference to all three.

No manufacturer of metal products will even consider production machining on an old manual Bridgeport mill, or cutting out parts with a handheld acetylene torch so why would you still have a weld shop full of nothing but manual welding machines? Is there anyone left out there that is still running production with acetylene welding or a stick welder of course not, but why?

It really comes down to a basic fundamental idea; you need to use the right tool for the job. If you need to drive a nail into a two by four stud you don't grab a pair of pliers do you? Wouldn't a hammer work better? If you need to put a screw into that same stud you don't grab a hammer, you pick up a screw driver and you also take the time and trouble to make sure that it is the right type and size as well. If you have to put in a couple hundred of those screws you'd grab an electric screw gun. The best reason to purchase a robot to do production welding is that it is the 'Right Tool for the Job'. That's it, it's not rocket science, it's just good common sense.

Lean Manufacturing is a term on everybody's minds today. The principles of lean production are derived from the Japanese manufacturing industry or more specifically the Toyota Production System. It is sometimes referred to as 'Toyotism' for that reason. The two main principles that it focuses on is smoothing out the flow and the elimination of waste.

A robotic welding system is one of the best ways to convert a conventional weld shop into the world of lean production. This is because the weld shop is usually where the work backs up and also where most of the overtime pay is accounted for in any metal manufacturer. Not only can a robot put out more work with the same size work force or less, they will speed up the production as well. Setting up a robotic welder requires a focus on the flow of material and completed work in and out of the robot cell, because of this aspect of robotic welding; having a one of these robots can be one of the best teachers of the lean manufacturing concepts. In other words buying a robot welder will force your weld shop to adopt the lean manufacturing ideals.

Many companies that are considering the purchase of their first robot approach the event with a high level of trepidation. This is because of the many myths associated with robots. These myths, like all myths, are untrue and need to be dispelled.

The worst of the lot is the assumption of the type of employee that it takes to be able to program a robot. It does NOT take a rocket scientist to learn the necessary skills, anyone of average intelligence can learn the process but a background in welding and some basic computer experience is a big plus. Another part of the same assumption is that the company will have to pay a six digit salary to hire a highly trained and experienced programmer. Actually, one of the best welders that you currently employ will probably make the best programmer for your robot; they are already familiar with the product that you are building and the expectations of your customers.

In most cases, if you offer a fifteen to twenty-five percent increase in a welder's pay you'll most likely get several volunteers out of your current bank of welders. This way all you will need to replace will be a regular welder.

You may assume that you will need a programmer for every robot that you put in, this assumption is totally false. If you only have a few robot cells the chances that all will need to be programmed at the same time are extremely low. A single programmer can usually do the programming for quite a few robots. However, I do recommend that you make sure that you have at least two programmers even if you have only one or two robots. If you have only one and he is absent for any reason; sickness, personal problems, quitting their job, or even death, that event could shut your robot division down just when you need it the most. I actually saw a case where the company only had one programmer and he went out for lunch one day and was killed in a car wreck. It took that company a few weeks to recover from that stupid mistake.

Most companies utilizing multiple robot cells have only two or three programmers while they have many workers that can operate a robot welder. A robotic welding cell operator only needs to be able to load the component parts, push the appropriate button, usually green, and then remove the finished product. The average time to train an operator is under an hour. I have joked about being able to train a reasonably intelligent chimpanzee or orangutan to operate a robot but it is actually not very far from the truth. Only when something goes wrong would the operator need to come and get the programmer and that does not happen often enough to need a programmer operating the robot.

Another myth is that only very high production runs can justify the expense of buying a welding robot. With the ease of programming the newer generations of robotic welders coupled with the shorter programming times needed, thanks to the highly interactive teaching pendants of today, much smaller runs can be very cost effective. This is especially true for repeat orders. Modern robot control units can store a great number of programs and if you have quick and easy mounting system for the fixtures, as I will describe in the body of this book, you can easily run several different jobs in a single shift. There have be many occasions when I have, on a single station, ran a dozen or more small quantity jobs in only eight hours. If your company deals with a lot of small parts where it is possible to place more than one fixture side by side on a single station, then it is also possible to run the jobs simultaneously through the use of a master call program. This flexibility associated with the modern welding robots is one of their biggest selling factors and the thing that makes them a resource that a manufacturer needs to keep up with today's unpredictable and sometimes explosive marketplace.

There are many reasons for company to buy MIG robots. One of the biggest reasons is the increased productivity that you can expect out of a robot. While it's true that a robot cannot weld any or much faster than its human counterpart, the speed which the robot can get from weld to weld can increase production by a factor of three or four. The greater number of welds that you have on your products, will determine exactly how much of an increase that you can expect to see. Also, when welding by hand you are either welding or unloading and loading the fixture but not both. A robot with two or three stations operating means that the robot is going to be welding while the operator is unloading and loading one of the other fixtures. It is possible for this one factor alone to double the productivity of a worker while keeping the robot in motion most of the work day.

Depending of course on the size and complexities of the jobs that you are running a three station robot with two operators can keep a robot moving the for the entire work day except for breaks and lunch because each operator can be unloading and reloading a station while the robot is welding on a third.

The repeatable accuracy of a robot welder is another reason for purchasing one. That accuracy not only allows you to give your customers exactly what they want but can save you a lot of money. When you compare robot welding to a manual welding operation you'll also find considerable savings just in the amount of wire consumed. A manual welder tends to make longer, wider welds than necessary just to make sure that there is enough of the weld on the product. With the consistency of the robotic welding this can be reduced to the exact amount of weld in each spot.

Most weld shops have at least a moderate turnover rate; this carries a cost for recruitment and/or training. All of the welders need to be highly skilled individuals with the higher pay that accompanies such positions. When you consider the rigorous training that it takes to produce a really good welder in conjunction with the relative scarcity of them in many regions of the country, the economic ramifications of these factors can be very significant indeed. Another thing that you need to consider is the fact that for several years there has been fewer welders being trained while many of the experienced welders are now approaching retirement. The consequences of this inadequate supply of needed personnel will sometimes include poor quality workmanship, the need to pay out unquoted overtime and most importantly the higher chance of customer dissatisfaction.

This gap is now beginning to be filled in by robotic welders. Over the years it has been claimed by many that robots are going to be stealing jobs from the people in the workforce, this is not true, and in many cases they are simply replacing a dwindling supply of trained workers or creating opportunities for the workers in the field to have a better career as a programmer of these robots.

If a company has eight or ten robots they can get by with only two or three programmers and the rest need only to be able to unload/load a fixture and push the green button. A robot that you have purchased does not quit, demand pay hikes or take vacations either. You also don't have to pay half of their social security or help on their health insurance. So when you take all of this into consideration buying a robot can be a lot cheaper in the long run than hiring another welder.

I teach welders how to program MIG welding robots and I wanted to find a good text book to help me with that endeavor but when I began searching for one I soon discovered that there aren't any on the market. That's when I decided to write this book, I want my students to have a book to help them in their studies and a reference for them to keep after I'm gone and they are on their own.

As much as possible this book is going to be written in general terms but when I am forced to use command names or controller key labels I'll be referring to the OTC Daihen Japanese robot arms. The reason for this is simply because these are the robots that I am the most familiar with and they are also the robots that I use to teach programming with. However, the basic principles of robotic programming are the same from robot to robot regardless of who the manufacturer was or where it was constructed. If you are using another brand of robot then you will just have to find the corresponding commands or keys that will perform the same functions with your robot.

I am also going to attempt to keep all of the instructions simple enough that just about anyone can learn to program a robot welder. While it does take someone with a moderate level of intelligence it doesn't require a genius level IQ to learn to program a robot. The two most beneficial prerequisites would be having a friendly relationship with computers and a background in welding. Neither of these is an absolute requirement but both would be extremely beneficial.

Although you will find a few tips in this book on troubleshooting welds, the instructional part of this book will mainly be dealing with the problems associated with the programming itself. Any references to welds in the book will be about carbon steel MIG welding, the one that I am the most familiar with. If you are going to be welding stainless steel, aluminum, or one of the many other alloys, you will need to adjust the instructions to take that into consideration.

The actual welding instructions and parameters will be different with every type metal that you work with and any of the different welding wires that are saturating the shops of today. This is why I am going to be concentrating my efforts on the programming part of the job, that stays pretty much the same no matter which metal and/or wire that you are using.

When you purchase a welding robot, the company that you bought it from will come and set it up for the type material that the robot will be welding. They will show you the weld parameters and how to adjust the settings to achieve the weld size and quality that you will need to satisfy your customers. All you will need to do at this point is learn how to fine tune the welds as you go along.

What they don't do is show you all of the finer points of programming, that's where I'm going to come in at. They will spend a few hours showing you how to write a basic program, how to use the keys on the teach pendant to move the robot around. They will also make sure that everything is working alright as well answer any questions that you can think of to ask, but then they'll leave and you'll be on your own.

The good news is that most of the people that try to teach themselves from this point can eventually learn most of what I'll be explaining in this book. The bad news is that a large percentage of these will end up crashing the robot several times over the first few weeks; some may even cause minor damage the robot before getting the hang of it.

Most of the companies that sell these robots also offer advanced training to the would-be programmer, they also charge a respectable price for this training as well. If you follow my simple instructions in this book, you will be able to teach yourself to program with the best of them and without damaging the robot at all. The only things that you will need other than my instructions is the will and desire to learn and time to practice a little, I'll take care of the rest.

Robotic Welding Plant Requirements

The most important requirement that comes with robotic welding is simply the consistency of the component parts. Robots are totally blind and actually rather stupid. They expect the seam that they are supposed to be welding to be in the exact same place every time. The tolerances of all material coming towards the robot must be held to a minimum; plus or minus only a few thousandths, the smaller the better. Some materials will nearly always give you trouble because of their mill's specifications. Angle and channel iron are two of the worse because the mill specifications say that as long as they are within an eighth of an inch they are well within their tolerances. That much difference will cause a robot welder to miss the seam entirely.

Inconsistent material size can literally be a nightmare with robotic welding. Any change in the material, whether it is too long, short, thick, or thin will move the seams that the robot is supposed to be welding.

The robot will not realize that and so it will continue to put the weld in the same place and if the seam moved then the weld will no longer be in the right place.

Robotic welding demands much closer attention to the quality and the manufacturing process of the piece parts than does manual welding. This may sometimes mean that parts that you had been sawing may need to be machined or some that have been burned out with a torch may now have to be cut with a laser. All of the parts being put into a robot fixture must be exactly the same. Look at it this way, a robot is really just a computer and what has always been true for computers is also true for the robots; Garbage in/ Garbage out.

When dealing with robotic welding systems the fixtures must be absolutely precise as well. This does not mean that the fixtures will need to hold the parts tighter, through the use of tack welding and by putting the welds on in the proper sequence you may find that you sometimes need fewer clamps than with hand welding. It does mean that the fixtures will have to keep the joints in exactly the same place on every unit.

These things are sometimes viewed as a disadvantage to robotic welding and I guess you could make a case for that, however if you take the long view, these requirements will cause you to turn out a much higher quality product which will in turn usually result in getting you more work. Regardless of how you look at it, getting more work for your company cannot be a bad thing.

While a skilled welder can easily compensate for differing dimensions and poor machining of the component parts being handed to him as well as sloppy and ill designed fixtures as well but a robot welder cannot.

Yet another thing that is very important to a robotic welding system is FLOW. The greater speed of production that is often associated with robotic welding cannot be achieved without being able to effectively have all of the component parts flowing smoothly towards the robot as well as the finished product flowing away from the robot area just as smoothly. There are two main factors that must be considered to create the smooth flow of parts coming towards the robot. The first thing that needs to be addressed is how you are going to get the parts from the previous department into the robotic welding area. The second thing is the placement, at the robot, of those component parts being loaded into the fixture. These parts need to be within easy reach of the operator and in the order that they are needed; the more that the operators are forced to think and move, the slower the parts will get loaded into the fixture. If the robot is sitting still waiting on its operator then it is not making you money. The robot is only paying for itself while it is working so the main focus should be about thinking of ways to keep the robot in motion as much as possible. If that means putting more operators at each robot then just grab another minimum wage employee and train them to load the fixture and push the green button. Your ultimate goal should be to keep the robot moving all of the time that the factory is open and running. To that end I have saw extra employees trained as operators just so they could be used to take the other operators place for restroom time, breaks and lunch time. Now I realize that this will add up to only an hour or so a day and that may not sound like much but most companies work at least two hundred and fifty days a year so that means that by recovering that time you will keep the robot working two hundred and fifty more hours per year. If your company, like most, work a forty hour week then that translates into just over six weeks' worth of robot time or man hours per year that you are going to gain, which is a significant increase in production.

Now, just for argument's sake, let say that you have four robots and you decide to do this, that adds up to a thousand man hours a year. I believe that would be well worth the effort and expense of training a low cost worker or two.

Choosing a Robot

There are a lot of choices today when shopping for a welding robot. Sadly, the biggest difference is sometimes more in the price tags than the quality of the product. Many robot manufactures spend far too much time and energy putting bells and whistles on their robots that you'll seldom use, many of which cost as much or more than the robots themselves.

Up until the turn of the new century the North American and German robotic equipment had led the field in robotic welding. But so far in the twenty-first century the leader in the welding industry has been the more cost effective, Asian built OTC Daihen robots. In my opinion they are the most user/programmer friendly robots on the market today. They're high quality robots that don't even have a scheduled preventive maintenance check up until they are three years old. I've never had a minute's trouble out of one of them and I have actually programmed some that were quite old.

The OTC Daihen robots are by far my personal favorite. Even though they are well suited for very large scale operations, because their flexibility, ease of programming and lower initial cost they are ideal for the smaller niche market welding facilities as well.

I have personally programmed many small jobs using these Japanese robots in less than thirty minutes. That combined with the ability to use relatively simple fixtures, many without using locking clamps to hold the parts in place, make them the perfect robot for many of the smaller, lower volume manufacturers.

So if you are simply researching the possibility of purchasing a robot or are already in the shopping around process, do take a close look at these jewels before you make your final decision. An OTC Daihen robot could not only save you a lot of money in the short run but with the long view as well.

Other Brands to consider

Fanuc – These robots are handled by the Lincoln Corporation and claim to be the World leader in industrial robotics. They are one of the most popular robotic systems being sold in the US. They are good robots but they are a little pricey for what you get and they are not the most user friendly to program. I think that the real reason that so many are sold is because they have latched onto the Lincoln brand name.

Kuka – These German robots are great if you need a robot with an extra-long reach or one capable of sheer brute force. They have about the largest, heaviest six axis robots on the market. They have only recently come out with a smaller robot but I haven't even gotten a chance to see one of those yet.

Motoman – These robots also claim to be the world leader in industrial robots but only the second largest in the Americas. Their largest claim to fame is their seven axis robot. I have seen this robot in action and I'll admit that there have been times when I would have loved to have that extra axis when trying to maneuver around a few of the fixtures that I have been handed. The biggest downside to these robots is their cost, what you get just doesn't balance out with the excessive out of pocket cost.

ABB – They offer robots with a wide range of size and configurations but they've spent too much time developing the hardware but their software leaves something to be desired. Programming these babies can be a nightmare.

Qirox – These German robots are supposed to be even more precise than most robots on the market. I've never had any personal experience with these machines so I don't have an opinion one way or the other. They are some of the most expensive robots on the market today. Taking into consideration just how precise all of the robotic systems are today I think that you'd be wasting your money on these babies.

Kawasaki – these robots are mostly light to medium duty. They have good software and are easier than many in their class to program. The joints of these robots move a little more in each direction before reaching the limits so they can actually reach more of their cell that a lot of other robots, this is a definite selling factor. They also cost a little more than some other popular name brands but may be worth it just from a maneuverability standpoint.

Qing Dao Metech – These are about the lowest cost robotic welding systems I've seen on the market today. I've never used one but I have seen one and they even look cheap. I have a feeling that this may be a case where you might get exactly what you pay for if you purchase one of these Chinese machines.

Setting up a Robot Welder

Your imagination is the only limit to the ways that you can set up a welding robot. The only real rule is that you will always want a way to safely set up the next job while the robot is welding; because the less time that the robot is sitting motionless waiting on you, the more productive it is and the more money it is making. Also, if you are an employee of the owner of the robot keeping the robot making money also has the added benefit of making your boss and the robot's owner much happier. Naturally, as the programmer, this can also affect your happiness as well.

There are three basic styles of setting up a robot:

Single station robots equipped with multi-station tables.

Multi station robot cells where the robot itself twists between two or three stations.

Track or gantry mounted robots that either moves between stations or around the perimeter of a single station.

There are advantages to each style but your choice needs to be determined by the size and type of product you are producing as well as how much room you have for the robot cell.

If you have a limited space and produce small products, the smallest set up that takes the least amount of floor space is a single station robot that is sitting behind a turntable. This is the perfect setup for a corner installation. The turntable is the key to this set up, and there are two requirements that must be met. First it is essential to have a divider in the middle so that you can be reloading the parts in one fixture while the robot is welding on the other one without the danger of flash burns. Second the turntable must have a solid lockdown in each position so that the fixtures are motionless and in the same exact same location every time. You are also not limited to a two station turntable. By using a little more floor space and a little larger turntable you can divide it up into three, four, or more stations. I saw one turntable that was split into six stations, cut up like a pie, each station having its own triangular compartment. This was done to give the finished product cooling time before removing it from the fixture.

Although I have seen a few of these turntable setups where the turntable was operated manually and even used one of them, the best setups have a motorized turntable and are controlled by the robot programming.

Another setup using a turntable is at the other end of the scale. In this case the product they were producing was very large with welds that were especially hard to get to. This fixture covered the entire turntable and this allowed the robot to weld on it from four different directions. They actually had two of these monster turntables sitting side by side with the robot in the middle. To be able to separate the two stations while welding there were two retractable curtains and the appropriate one would drop down behind the robot just before it began welding. The curtains were controlled by a subroutine triggered by the choice of stations.

A high-tech motorized version of the turntable principle is the positioner. The beauty of the positioner lies in the fact that it doesn't need to have the lockdown positions. This handy tool is controlled by the robot program and can stop turning anywhere along the full three hundred sixty degrees of motion and as many times as necessary. They can be used alone either horizontally or vertically, and can also be used in combination with a freewheeling partner to create a rotisserie styled framework for your longer fixtures. Positioners can also be used for continuous motion, at any speed, where the robot will be welding a continuous weld around a product such as a pipe or tank. Positioners are one accessory that could be used to make nearly every robot more efficient and compared to their relatively low cost they can be a very wise investment. In many cases they can completely eliminate the need for multiple fixtures and two stage programs allowing the product to be completely welded on a single run using only a single station.

The most common set up for MIG welding robots is a multi-station cell. Most are square cells with either two or three stations. I recommend that you avoid the cells that have two stations on a single side of the cell, I have used them in the past and they have an inherent flaw that cannot be worked around. No matter how a robot station is set up the station needs to be centered along the center axis of the robot's base. Neither station of these flawed cells is lined up properly with the robot arm which will likely cause many problems and headaches for the programmer. These robots loose a great amount of possible reach because of the need to navigate the offset.

If the products that you build are flat and the fixtures are designed where the robot will not need to thread the torch in and out of tight spaces then this type setup may not give you problems. Fixtures or products that the robot will have to either reach around or through will give the programmers more difficulties as they write the programs.

If you must deal with one of these cells the only thing you can do to eliminate some of that problem is to angle all of the fixtures toward the center line of the robot. Even though this remedy will not correct all of the problems associated with this design, I think you'll find that it is the best that you'll be able to do.

The best setup for a square cell is to have three stations on the front three sides of the robot leaving the back side for the access door, the controlling computer, and the drum of welding wire. Some of these cells are not designed as well as they could be either, the tables are either too close to the robot or too far away. All the stations also need to be at an equal distance from the robot so that you won't end up with fixtures that can be run on only one station of the cell. I've saw a few robot cells that were built with one table as much as twelve inches further away from the robot than the other two and yet none of them were in the right place. Many times the one table will be too far away from the robot while the other two are far too close to it. Ideally you'll want to be able to reach the two farthest corners of each station's table with the nozzle held vertical and the joints NOT completely stretched out but close.

Another type setup that can work well for the larger products is a single station robot combined with a two station sliding table built on a motorized track. These are wired up so that the table shuttles when you push the start button and when the table locks into position a curtain or wall drops down between the stations just before the robot begins to move. The neatest one that I have saw using this technique was built through a wall so that once the table had shuttled through and the air actuated door dropped into place, the fixture that the robot was welding on was in one room while the other, ready to be unloaded and reloaded with component parts, was in the other room. This setup completely eliminated the chance of the operator getting their eyes flash burned. The door between the rooms was also wired into the system so that if it was opened manually while the robot was welding it acted as an emergency stop and would shut the robot down instantly.

If the products that your company builds are too massive for a conventional robot cell you may think that robot welding is out of the question but it's not. There is another type of set up that is made just for your type of situation, where the robot is mounted on a motorized, track based platform. The tracks are usually either horseshoe shaped or as an incomplete circle. While the robot cannot travel completely around the work station, the reach of the arm from both ends of the track will usually allow the robot access to the entire product. These operate very similar to the stations that use the positioners only instead of the work moving it is the robot that is mobile. With this set up you will program the robot to go to a specific location on the track before starting. It will make all of the welds that can be reached from the starting position before moving to another location and beginning the second round of welds. This can be repeated as many times as necessary until all possible welds have been completed.

Another solution that is commonly used when building large products is the gantry system. These gantries can be mounted in a great variety of positions; upright on the floor, inverted overhead, a horizontal or vertical on the wall or even frame mounted. I saw one robot that was mounted on a double overhead gantry system that was held aloft on a track mounted, motorized frame. The gantry that the robot was mounted on was vertical allowing the robot to be raised and lowered through a twenty-four inch range of motion. This gantry was mounted to a horizontal gantry that had a range of over eight feet. The frame was mounted on a track that allowed the gantries to travel twenty-two feet across the floor. The frame and both of the gantries were controlled by the robot programming. The thing that made this set up so unique was that the robot itself could be moved in all three directions throughout its own X, Y and Z axis pattern. This made it one of the most versatile robot systems that I have ever saw, I just hate that I didn't get the opportunity to sink my teeth into programming this baby.

On these track or gantry mounted setups you are also not limited to having a single robot per cell either; there can be multiple robots each with their own section of the robotic cell.

Another track mounted set up that I have saw in operation uses a straight track that is inside of a large rectangular cell with two work stations on each of the long sides. This robot has the ability to run four different jobs at once. When I left there that robot cell was working so well for them that they were talking about purchasing another one just like it.

As I said before, when you are designing a robotic cell you are only limited by the depth of your imagination. There is no best way to set up and use a welding robot, their beauty lies in their adaptability. I saw a video of one that was mounted upside down on base that was attached to the ceiling by screw jacks so that it could raise and lower itself up to eight feet with commands that were wrote into the program. I would love to get my hands onto that setup too, I can envision many possibilities that I would love to try out on a setup like that one. That one in particular was being used to weld the inside of stainless steel tanks while all of the parts were held together from the outside. After it got through the tank was then shuttled over to another robot that welded the outside of the tank.

If you have a two or three station cell you'll need some sort of table, frame or a rotisserie styled positioner at each station to mount the fixtures to. If you build a great variety of products you may want to have a different styled setup at each of the stations. I've saw some that were set up with a plain table with a grid of threaded holes at one station, a table with a small positioner at another and at the third there would be a large rotisserie style floor mounted positioner/tailstock combination. With this styled cell there are few jobs that will fit inside of the cell that the robot cannot handle on one of its three stations.

You will need some sort of grid to attach the fixtures with. I personally prefer holes that have been threaded for half inch bolts in a three inch by three inch pattern covering the entire table or frame. I have also found that it is best to plug the unused holes with silicon cone plugs that are commonly used in powder coat painting operations. These plugs will keep the weld spatter, dust, dirt, or anything else from clogging up or damaging the threads but are easily able to withstand the heat of the spatter without melting. The grid should be labeled so that you can bolt the fixture down to the exact same spot every time. I use numbers for the rows and letters for the columns. However you choose to label yours you should stencil the grid location onto the fixture and make a note of it in a comment line of the program.

Regardless of which style set up you are using there are certain considerations that you'll need to look at. One of the most crucial things that you will need to consider is the floor space. Beyond the space that the robot takes up you'll need easy forklift access to the area were you will have the wire drum. Although they vary greatly in size and weight the lightest is heavier than most people want to be moving around by hand.

If your robot has a three station cell then you'll need three separate staging areas for the component parts plus three areas that are designated for the finished products. On top of this you'll also need to retain enough clear space so that you will be able to move safely from station to station. The amount of space that you'll actually need will be determined by the size of product you'll be building and the amount of component parts you'll be using to put them together. The amount of area needed can sometimes be reduced by creating a flow of parts toward the robot as well as a stream of products flowing away from the robot, this ties in with the lean manufacturing concept that I talked about earlier. This is actually the most efficient type of setup for a welding robot. The best way to achieve this flow and keep it operating smoothly is by utilizing another lean manufacturing concept; the 'Water Spider'.

The Water Spider is the whole key to the concept of lean manufacturing. When first introduced to the basic duties of the water spiders there is sometime a misconception that their job is a simple material handler or just a 'Go Fetch' laborer, the individuals that make this assumption have missed the point entirely. A Water Spider needs to be thoroughly familiar with the materials and tools associated with the work stations that they are servicing. They must also be knowledgeable and conversant with the processes involved as well as the product being built. Water Spiders are by no means a minimum wage employee, it can even be said to be a management training program, many water spiders have eventually moved on to become excellent supervisors.

When dealing with welding robots the water spiders must also be familiar with all of the consumables that the robot uses and keep a supply handy for the operator and programmers.

Along with keeping a steady flow to and from the robot, another one of the water spider's duties is to make sure that the parts and fixture for the next job is waiting close at hand. The best water spiders will also be able to take the last fixture off of the robot and install the next one. They also must be at least as familiar with the fixtures as the operator so that they will know in what order that the parts go into the fixture. This will enable the water spider to know the best way to arrange the parts in the staging area.

The good news is that after you get the water spider concept up and running for your robot, you will see other places in your facility that could make excellent use of a water spider in their areas as well. If you manage to get water spiders servicing all of the departments in your business you will be well on your way down the road to becoming a lean manufacturing facility, your production will increase, your waste will decrease and you will probably end up getting a lot of your competitor's customers to call your own. That's a lot of benefits from the seemingly simple decision to purchase a welding robot, but it is what many new robot owners have experienced.

Each application and type of product will be different and you'll have to decide which type of setup will work best for your business. If your company builds a variety of different products you may find that you have to have different styled setups for different robots, in fact most manufacturers that purchase more than one robot will end up with different setups for each one. This can help you bid on and build an even greater variety of jobs in the future.

Another thing that you'll need to take into consideration is adequate ventilation. How much you will actually need depends on several factors such as what materials that you will be welding, the height of the ceiling and how well your facility is sealed. If you are welding steel then you may be welding oily metal and dealing with smoke whereas aluminum or stainless steel is always very clean. The metal that requires the most ventilation is quite probably galvanized steel, it produces a poisonous gas as you weld it.

Lighting is for the programmer, the robot actually needs none. I have seen what is called lights off operations where a room with over a hundred robots; MIG and spot welding along with material handling robots all working together on an assembly line with no human workers at all. The entire operation is done it total darkness. If there is a problem with one of the robots the line comes to a stop and a single light comes on over the robot that is having the problem as it signals for help. The programmer, on the other hand, needs as much light as possible, from several different directions. Without sufficient light the programmer will not be able to get the welds programmed in the right place.

Two things that I like to have close to every robot station are an air hose and electricity. Of the two, I think that the air hose is the most important because the operator will need to keep the fixture blew off using a good air nozzle. If the fixture is not kept clean then the component parts will not be in the right place causing the weld to be put in the wrong place on the product. It only takes one spatter ball to throw everything off. A blind robot puts the weld in the same place every time but that repeatability only works if the seam remains in the exact same place every time. The main reason that I like to keep electricity close at hand is because there always seems to be a need to use either electric tools or extra lighting and I'd rather have it and not need it than need it and not have it.

Unless the robot is in an air conditioned environment the operator is likely to need a fan or fans during the summer. Be careful not to have the air blowing inside of the robot cell. The shielding gas can easily be blown away causing porosity in the welds.

MIG welding robots have a wire lead as well as the power lead. Both of these will need support and because of the energetic movements of the robots this will need to be done through the use of a retractable tool balancer. I've saw everything from springs to bungee cords used to support the leads and get them off of the floor but none work as well as the balancers. The tension of a balancer can be adjusted so that you can get just the right amount of support and they don't wear out and break as easily as the other options. These are not terribly expensive and can be worth their weight in gold.

Robotic welding involves very precise tolerances and even minute variations in the wire feeding can result in unacceptable welds. It is very important to choose a wire that will pull easily through the drive rolls of the wire feeder and glides effortlessly through the liners. The worst decision that I've yet saw by a company that is running robotic welders is to try and save a buck or two on the wire. Most of the time when you choose a lower cost wire that is produced by one of the aftermarket sources; you actually do get exactly what you pay for. This is one way that saving pennies can end up costing you hundreds or thousands in lost revenue down the road.

Another consideration that is peculiar to robotic welding systems is the arc starting performance of the wire. This is another area where those low cost alternatives fall short. A wire that produces either untrustworthy or unpredictable arc starts can easily cancel out the benefits of the implementing of a robotic welder because of the downtime and/or manual reworking of the welds.

For most of the MIG robotic applications that I have personally dealt with metal cored wires have produced the best results. Metal cored wire flows onto the product smoother and produces a high quality weld with a good travel speed especially if you have to deal with mill scale on the metal that you are working with. The metal cored wires also produce less weld spatter which reduces the amount of post-welding clean up and also the frequency that the nozzle and contact tip will need to be cleaned and/or replaced.

Another benefit associated with the metal cored wires is that it reduces or eliminates the sub-surface porosity that is sometimes caused by the wrong wire extension length or changes in the angle of attack of the welding torch. In other words, metal cored wires can be much more forgiving and that is important, especially if the products you are building have some hard-to-get-to welds where you can't hold the optimum positions, directions and angles when getting the welding torch to the joint.

Another area that it is not wise to go with the low-cost alternatives is with the consumables. Because robotic welding sometimes involves numerous and frequent tack welds it's very important to choose consumables that reduce arc start failures and/or frequent changes which can and will produce avoidable downtimes. High quality consumables can save you a lot of money in the long run because if you save fifty percent on contact tips but you have to replace them twice as often then you are actually losing money.

Robot Safety

Too much cannot be said about safety when dealing with robot arms. Robots can be very dangerous machines, that is why they should always be separated from the operator/programmer, especially while the robot is in either the playback or run modes. They are very fast, extremely powerful, and they do not have a conscience at all. There are many different ways to achieve this separation, which you decide to use is up to you but please use one of these or devise one of your own, robots can and will KILL.

If you have a closed cell, all of the doors need to have sensors engaged before the robot can begin its operation. The walk-in door is always a manual door with a solid latch but the station doors are usually controlled with a subroutine in the basic programming that is put into place when the robot is installed in the cell. These doors can be opened and closed manually but they are usually operated automatically through the use of air actuators or hydraulic cylinders but I've also saw cable, belt, chain and gear driven doors operated through the use of servo motors much like the ones that power the robot itself.

If you have a robot that cannot be enclosed inside of a secure cell then you must use another method to protect the operator/programmer. The two most common methods are through the use of either light curtains or pressure pads. The light curtains use laser beams projected from a sending unit to a mirror which reflects it to another mirror and finally into a receiver creating a three sided box. The moment one of the beams is broken it causes the robot to be shut down instantly. Pressure pads work in the same manner; you line the border of the safety zone with the pads and then the slightest pressure will cause the robot to shut down. Both methods will keep anyone from stepping within reach of the robot while it is in operation.

If you have a robot cell with a floor mounted positioner many of these setups have a garage styled roll up door that is far too easy to walk through. If you have this style of station I would suggest that you also invest in an alternate safety barrier such as the afore mentioned light curtain or pressure pads to insure that there is no one inside when the run is started. Your first concern should always be to prevent anyone from being in close proximity to the arm when the robot begins to move.

It is also possible to get hurt while programming a robot as well, especially with the older ones. Many of the newer models have a lot more safety features built into them. Take the OTC Daihen robots for example, the teach pendant of the older models have a dead man switch that you must hold in while the robot is in motion. The idea was that if you got pinned in between the robot arm and the fixture you would release the switch which would keep you from becoming a dead man. The flaw with that scenario is that many people in that situation will actually tighten their grip instead of releasing it. The designers changed the way the dead man switch works on the newer models to solve this problem. On these new pendants you must hold the switch only to the halfway point, a narrow 'sweet spot' that you have to keep it in to move the robot, this means that either releasing or tightening your grip will shut the motors off.

All that being said; no matter how many safety features that are built into the particular robot that you are programming, DO NOT EVER put yourself in harm's way. Always think about where you are relative to the robot, the fixture, the product, the wall of the cell, etc… NEVER allow yourself to be in a position where the robot could hurt you if something went wrong. The base joint of a robot is so strong and moves so quickly that it could cut you in half in less time than it takes to think about it.

You may have read or heard that the robot has a safety pressure limit that will shut the motors down before any serious damage can occur. What you may not realize is that those measures were put into place to protect the robot not the operator. Think of a robot the same way as you would a semi-automatic pistol, it doesn't matter how many safeties the pistol has on it, it just isn't a good idea to stick it in your belt and carry it around with a live shell in the chamber.

The dead man switches that are built into most of the robot control units or teach pendants are made, as the name implies, to keep the programmer alive. These switches will not necessarily prevent you from serious injuries such as severe bruising and/or broken bones.

Personally, I have never fully trusted those dead man switches to keep me alive either, there have been far too many times in my life that I have flipped a switch and nothing happened. Any switch can fail, even those that cannot. Never trust anything but your good sense when your life is at stake, don't take chances with anything as potentially dangerous as a robot.

I have actually saw a programmer wedge a dime in beside the old style dead man switch on a teach pendant to lock it on so they wouldn't have to hold it constantly. I saw another one bypass the safety lock by going in over the table while the robot was ready to run. Yet another programmer that happened to be programming one of the older model OTC Daihen robots that didn't have a Check Weld function had someone lock the walk-in door, turn the motors back on while he was in there with the robot. Then he had his helper began a run just so that he could observe the robot as it made one of the welds. Luckily none of these individuals got themselves hurt but they could have been hurt bad, possibly even killed.

They don't make a safety lock that is so foolproof that it cannot be bypassed by someone that is determined to get around it but I would never risk my life doing such a thing. Actions like those go far beyond being careless or even reckless, that's just pure stupidity.

Using the Torch Alignment Tool

All welding robots come with a torch alignment tool. Many manufactures call these handy devices a J tool, I suspect that this is because most of them are shaped roughly like that particular letter of the alphabet. They usually have at least two holes to bolt it on with, screws for you to attach it using threaded holes in the last axis of the robot arm and a pointed tip pointing up towards the same axis. It also comes with a pointed tip that is used to replace the contact tip with.

The principle is very simple; once the contact tip is replaced and the torch alignment tool is bolted on securely the two points should be in perfect alignment with the two points just barely touching each other as shown in the illustration below:

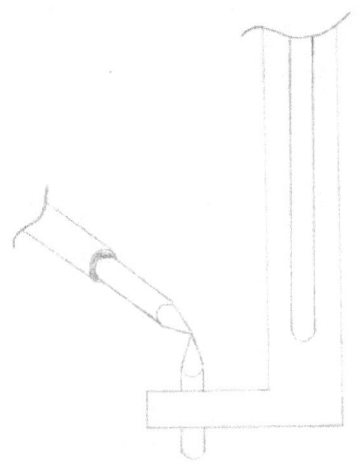

The points are not straight in line with each other; this is because it is a generic tool that is made to be used on all of the different torch angles. I understand the principle behind this but I don't understand why they couldn't have made a different tool for each type of torch or at the even two or three variations so that a company could get one that was closer to the proper angle for their particular torch. At the very least I feel that they should have made the 'Generic' tool at an angle that would be considered a midrange angle. As far as I know each of the different manufacturer's alignment tool is set up for a straight torch which is probably the least used torch that any of them offer, that makes absolutely no sense at all. As it is, you just have to do the best you can with what they give you.

If it happens that the two do not meet each other as they should you will have to do some tweaking. There are always three directions of adjusting the torch that corresponds directly with the directions of movement for the robot; X, Y, and Z.

Usually two sets of adjustment screws will have slots around an inch long while the other adjustment allows the neck of the torch to twist.

Making the adjustments can be a bit tricky; every time that you think that you have it right, the points will move slightly as you tighten the screws. You'll find that you will have to allow for this unexpected movement, it can be a bit tricky but in this case trial and failure is the only reliable method and persistence is definitely the only key to success.

Note: Only manipulate one adjustment at a time, get it as close as possible and then tighten that one securely before loosening the next adjustment slots. Keep doing this just as many times as you need to until the two points meet perfectly.

Using this tool regularly will keep the robot running true to the programming. Sadly, most programmers only put this tool on if they slam the nozzle onto the table or the fixture and then only if the robot begins missing its programmed marks drastically. It's been my experience that if you wait until this point you have already done some programming with the torch out of alignment. If this is the case you will eventually have to go back through these programs and correct them once you have realigned the torch.

The fact is these adjustments do slip over time regardless of how tight you manage to get them. If it's been a long time since you used this tool and checked the alignment then it is probably going to cause you a lot of extra work when you finally get around to it or are forced to do it because of a crash. I know one programmer that went over a year without checking and truing the alignment before he forgot that the robot was set in its highest speed and slammed the torch into the table. This resulted in the torch being jolted out of position by over a half inch. After realigning the torch the programmer finished writing the program that he was working on and thought that everything was back on track, until he loaded the program for the next job. This was a repeat order that he'd just ran three weeks before and it had ran perfect then. He didn't check the program and when the robot ran the first cycle he found that the welds had missed the seams by nearly a quarter of an inch.

After that he discovered that all of the programs that had been written in the past six months were off significantly but with the more recent programs being the farthest out of position. That's when he realized that the torch had been slipping out of alignment slowly but surely for over half a year.

The reason that I know this so very intimately is because this programmer that learned this lesson by an extremely harsh and bitter method is someone that I know very well, I was this programmer. This was a hard lesson to swallow but one that I learned well, I've never let this one happen to me again. This is also the incident that caused me to create what I call the 'Check Point Program' that I will discuss in detail in another chapter. This simple little program that only takes a minute to run will show you if the alignment needs to be checked.

It is much better if you check the alignment at least once a month or more to prevent this kind of incident from happening to you; an ounce of prevention really IS worth more than a pound of cure. Pick a time such as the first Monday of every month and make this into a preventative maintenance regimen, it will save you a lot of work in the long run.

All of the robot welders that I have ever dealt with also have a check program preloaded into the list of available programs. Running this program will put all of the robot's joints back into a factory set position where all of the joints will have an arrow lined up with a checking mark. I usually run this program before checking the torch alignment. So far I have never had this to be mis-aligned but once and the owner called the company that sold them the robot and they sent their own repairman out to fix the problem. It turned out that one of the belts was worn out and had to be replaced. This is not something that is the programmer's responsibility to fix as is the torch alignment, but it is the programmer's responsibility to find the problem.

If you are programming an OTC Daihen robot this program is usually numbered 9999.

First Steps in Programming

If you have never programmed a welding robot then this is going to be one of the most important chapters in this book. You'll need to read through and follow the instructions in the order that I have written them. Contained in this chapter is the first things that you will need to learn as well as the first exercises that you will need to perform on your way to becoming a robot programmer. It is important that you successfully complete each assignment before moving ahead to the next exercise because they are going to be increasingly difficult as you go through the chapter.

To become a welding robot programmer, the first thing that a robot programmer must do is to acquaint themselves with the robot's movements. This can only be accomplished by playing with the robot, just by taking the teach pendant and using the movement keys to guide the tip of the arm around in the useable three dimensional space that belongs to the robot. A programmer must be able to place the tip in every reachable point within that space and in all the possible positions at each of those points. The illustration below shows a diagram of the useable space of a three station robot cell.

You'll notice that some of the apparent useable space falls outside of the cell in front of each table; this is done so that the robot can reach further into the corners of the tables of each station.

Another thing that is noticeably absent from the diagram are flash curtains. This is because there are a great many different types, styles, and sizes to choose from, so many that most of the manufacturers of these cells don't even offer any as an accessory.

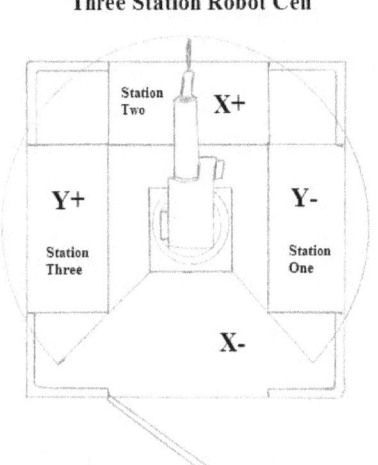

Three Station Robot Cell

One of the first things that a potential programmer must do is to get acquainted with this three dimensional space or spaces; I actually consider each of the stations as a separate cell. This is an important first step in learning how to position the robot's torch. This can be likened to a baby learning to grasp something and bring it towards themselves. Before a baby can learn to grab and pick up an object they must first learn how to get their hand to the object. The very same principle is applied to the robotic welder, if you can't get the tip and nozzle into the correct position then the robot cannot make the weld.

The very first position you must establish is the robot's home base. The location of that home base should be determined by several different factors. If it is feasible with the configuration of the robot's cell I always try to choose a home base within easy reach of the operator so that the nozzle and/or tip can be examined and changed if needed. If this is not a viable option then you will need to add a cleaning point in most of your programs where the nozzle will be within easy reach of the operator for checking, cleaning or replacing the consumables. If this is the case you may find that the best location for your home base is one where the arm will be poised to swing past or dodge its curtains or where it will have easy access to any of the stations. The robot will always get to the station in front of the home position with fewer moves, for this reason if I'm going to be using one of the stations more than the others I'll try to have it as the one that is in front of the home base. Most programmers of three station robots usually create their home base with the robot poised in front of the middle station. One distinct advantage that you gain by doing it this way is the fact that although they will still have less moves to get to the middle station they will also be in an equal posture relative to the other two stations as well.

After establishing a common home base for the robot I like to create a simple check point program. Choose a section on one of the station's tables and put a good center punch mark in it, about an eighth of an inch deep. The best table to use is the one that is in front of the home base position because there will be fewer moves involves. Make sure that it is in an area with good lighting so that it can be seen easily. You should inch out enough of the welding wire so that the length sticking out beyond the nozzle can be trimmed to exactly five eighths of an inch beyond the nozzle when creating and using the Checkpoint program. Now simply maneuver the robot so that the tip of the wire is exactly touching the center point at the bottom of the mark as shown in the illustration below. The end of the nozzle should be a half of an inch above the metal. Be sure to have the preceding point directly above the check point to prevent the wire from touching and bending on the way down into the mark. Record a stop function into the program so that the door will open giving you good visible access to the mark. Finish the program by returning the robot to its home position and ending the program.

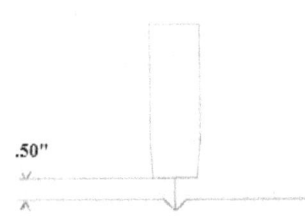

Once this program is in place you can trim the wire to five eighths of an inch and run the program so that you can check to see if everything is in alignment. Run this test often and you'll never have to worry about the need to go back and rewrite programs after the alignment has been corrected. This is a simple preventative maintenance task that is easier and quicker than performing the alignment check and can save you much time and effort later on. If the robot fails this test then it's time for you to get out the torch alignment tool, put it on the robot torch and preform the real alignment check.

If the stations of your robot have a grid of holes to locate and attach the fixtures with that's great, if it doesn't you need to come up with a way of creating a grid. Grids enable a programmer to place a fixture at the exact same place every time, this much more than a convenience, it is essential so that you will be able to return the fixture to the exact same location each and every time. There are several ways to mark a grid but my favorite is to use numbers on the rows and letters on the columns.

After these preliminary tasks are out of the way you'll need to begin practicing moving the robot. The best way that I have found to do this is to set up an obstacle course to maneuver the robot arm in between. Use objects that are lightweight and easy to move so that when, not if, you hit them the objects will just slide out of the way without any resistance or damaging anything.

Some of my favorites for this exercise are empty soda cans and bottles along with various sizes and shapes of boxes. Set up at least ten or twelve of these at different locations around the table and be sure to mark around them with a felt tipped marker or silver pencil so that when you move them with the torch or wire you'll be able to replace them in the exact same spot.

Next, lower the torch to a point about two or three inches above the table. Pick a spot that is an equal distance from and in between four of the objects and record the position. After that you'll need to move the robot between two of the objects into another mid-point and record it. Repeat this until you have moved the torch between most or all of the objects and recorded at least twice as many points as you have objects in your impromptu obstacle course, be sure to keep the nozzle perpendicular to the table and also to record a few points on the outside of the pattern as well. During this session never lower the robot any closer to the table so there won't be any danger of collision with anything except the movable objects that you are using for the obstacle course during the playback mode. After maneuvering through the course bring the robot back up to the home position and record and end to the program.

Place the program that you have just created into the proper station and switch the robot into playback mode. I strongly suggest that you override the speed down to thirty percent to further insure the safety of the robot. If you happen to be using an OTC Daihen robot this is accomplished by pressing and holding the Enable key while in playback mode which will change two of the function keys and allow you to override the programmed speed for the test.

Now all you need to do is run the program, don't be surprised if all or most of the containers go flying off of the table, which happens to most students on the first try. If that doesn't happen then you should give yourself a pat on the back, if it does, then you can set them back up using the marks that you traced around the objects. After getting the objects back into place on the table you need to jog the robot through the program until you find the point that caused the dominoes to begin falling. There is a real good chance that you will find that it happened because you forgot to record one of the points that you moved the robot to while writing the program, that's actually a pretty common occurrence.

After you have successfully written an obstacle course program that the robot can run through without disturbing any of the objects then you need to create one on each of the other stations. As you create these other obstacle course programs you will find that moving the robot while standing in front of each of the other stations is a lot different and that is why you need to do it. Because you will be writing programs on all of the robot's stations you will need to be familiar with the movements associated with each station.

One of the most important things to learn during these practice secessions is how to move the robot around in the three dimensional space that the robot maneuvers inside of. This more commonly referred to as the X, Y, Z coordinates. This is the how the robot knows where it is at and enables it to return to that exact point within the cell every time the program calls for it.

If you have a three station cell with curtains, which is one of the most common setups for enclosed cells, the robot actually has three separate sections of this three dimensional space to become familiar with. As you saw with the last exercise the X and Y coordinates switch positions as you move from station to station and the positive and negative directions reverse between stations as well. It helps me to think of each station as a separate cell because of these differences.

Inside the station that is in front of the middle station X+ will be towards you while Y- is to your left and Y+ is to your right. On one of the other stations X+ is to your left while Y+ is toward you and Y- is away from you. The moves on the opposite station will be completely reversed with X+ to your right, Y- will be towards you and Y+ will be away from you. So if you look at it from this perspective, they really are like having three totally different cells with completely different movements. There will be many times when you will press the wrong movement key because of these differences so don't be surprised when it happens.

Unless it has been changed by a former owner of the robot, almost all welding robots have the same coordinates. If you are standing in front of the robot, that is opposite of the side of the robot that the cables are attached to, then +X should be toward you and –X away from you. +Y should be to your right while =Y will be to your left. +Z should always be up and –Z down. The only time that may be reversed is on a robot that is mounted either on the wall or overhead. If it's your first time on a used robot then never assume that any coordinates are as they should be because they can be changed, always check it with the robot in one of the slower speeds to make sure.

There are exceptions to every rule and this one is no different. There are a few robots like the Nachi for one that the Y axis is from front and back while the X axis is to the right and left.

To move the robot in the right direction every time you'll need to have this pattern in your head. In the first few days and weeks you will either find yourself hesitating so you can picture the pattern in your mind before moving the robot or pushing the wrong button and moving the robot in an unexpected direction. Don't let this frustrate you, before long you'll have it down and it will be there forever. After that it will only be a matter of remembering which station of the robot you are working on. With all of these years now under my belt I still find myself looking up from the teach pendant once in a while to verify which station that I am standing in front of. However, there is always going to be the possibility of moving the robot in the wrong direction, which is why you will constantly need to keep a close check on what speed you have the robot set on. Nothing will ruin your day like slamming the robot into the table or the fixture at full speed.

Once you have played around with the robot movements enough that you can safely maneuver the robot around each station of the cell then it is time for the next task, creating station templates for each station. This is necessary for all multi-station robot cells but if yours, like most, has flash curtains that the robot will have to dodge before starting each run, then these templates become even more important.

Station templates are great time savers because by making a copy of the appropriate template and then using the copy as a platform to build the program on will give you a head start by eliminating the time and effort that you would have spent programming those lines and robot movements that you will need at the beginning of each program.

Station templates are programs and as such need to start with comment lines explaining what they are. Comment lines are strictly for the programmer and/or operators and are ignored by the robot. With most robots the first comment line is used in the program list as the description so you need to make sure that you use an easily recognizable short comment. If you can't make it shorter than the number of characters that can be displayed in the program list then make sure that the first ten to twelve characters contains enough information that you can easily recognize the program from the list. With a three station robot the first comment line for the station one template should oddly enough read something like "ST 1 Template".

The next comment line should show the grid location where the fixtures are to be attached to the station's table or frame. An example of that line is "Grid location (X-X) (X-X)".

After those comment lines the first actual programming line should be the common home base you have already established. One of the most common errors made by beginning programmers is to move the robot toward the ending point of the first movement without first recording the home base while the robot was still sitting in the home position. Once you record the home position then the next thing that you will need to consider is how you are going to get around the curtain or any other obstacle in the robot's path on the way to the station containing the job. If you are going to be dodging a curtain the first move needs to be pulling back and raising the robot arm to a position where it will miss the curtain as it swings past it. The next position should be in line with the closest approach to the curtain but several inches away from it; this will enable the robot arm to swing past the curtain with good clearance. If you were not able to create a home base in a position that would enable you to check and clean the nozzle then the next point you will need to add to the program in each template is one that puts the nozzle within easy reach of the operator when running that station.

The nozzle should be positioned where it will be pointing towards the operator while their finger remains on the start button and where it can be seen with the door still shut if possible. This will speed up the running of the program because if the nozzle and contact tip is clean and in good shape the button could simply be pushed again even before the door opens completely or at all.

Finally you need to get the robot into a position where it will be able to safely dive in towards the first weld but still be above the tallest fixture that you are likely to get handed.

This is usually a good place to end the station template because all of these movements and points will be needed at the beginning of each and every program for that particular station, but any points beyond this position will be different for each particular job or project.

Now you need to repeat the previous steps for the other stations until you have created a station template for each of the other stations of the robot cell.

Once the station templates are saved it is time to start practicing writing programs. I suggest that you practice programming around something that is very lightweight and moveable such as the objects that you used in the obstacle course from the first exercise. This is so that you don't have to worry about damaging anything until you get the hang of moving the robot around in close quarters.

Choose which station you wish to use and begin by making a copy of that station's template.

Note: Always make a copy of the template and use the copy as a base to write the rest of the program on. Never bring up the template and begin adding to it because you will need it for the next programming job. This is the first and one of the most important rules a programmer needs to live by; ALWAYS work with a copy, NEVER with the original. This applies to editing as well as when you are writing new programs.

Once you open up the copy of the station's template that you just created you'll need to put the robot into edit mode and by using the arrow keys, move down to the first comment line and then over to highlight the comment. Now bring up the soft keyboard by using the Enable and Edit keys together and switch that first comment line to a name that you will easily recognize and then click on 'complete' to exit the keyboard. However, without a real fixture to use just leave the second comment line for the grid location in its generic form.

Bring the robot out of edit mode and put it into teach mode and then check forward through the program until you reach the point that you recorded after the cleaning position that is at the end of the template. While the robots is in the inspection/cleaning position always make sure that you have a new contact tip installed and the wire is protruding from the tip the correct amount. A good rule of thumb is that if you are going to be welding at or below two hundred and fifty amps the wire should be one half to five-eighths of an inch past the end of the contact tip. For the hotter welds with settings over two hundred and fifty amps this should be increased by approximately one eighth of an inch.

The next point in the program should be at a position just above the beginning of the first weld. If possible, the best position is one where the nozzle can dive straight in towards where the arc start point will be located. I personally like to have it about two inches away if possible. The next point will be the start of the weld. All of the previous programmed points were approximate but this one will need to be a very precise location. This is usually the point where I break out the flashlight because regardless of how much light is shining above the robot there never seems to be enough to be sure that the end of the welding wire is in the exact location that I want it to be. If you happen to wear a cap and you can find one of the lights that are made to be clamped onto the brim, those can be real handy because it will free up both hands for handling the teach pendant.

All open air movements only need to be, as I call them, eyeball close, even those when navigating around the fixture and/or part. The two points of a line or the three plus points of a circle or arc are the only points in a program that has to be absolutely exact. This precision needs to include not only the location of the tip of the welding wire but the angle of attack as well.

Because of the need to be so precise always double check, before moving the torch into position to start the weld, that the wire extension length is correct. Many times as you twist and swing the arm around the cell it will directly affect the wire extension length sometimes as much as an eighth of an inch or more. Needless to say this will adversely affect any weld you program after this happens so you need to keep a close check on this.

Note: All the way through the programming phase I always keep a diligent check on the wire extension length especially just before I get to the start of each weld. I do this because as the torch and arm flex it bends and twists the liner that is between the drive spools and the tip which causes the extension length to change.

Make sure that the robot is set on the proper movement type and one of the lower speeds then slowly move into the position for the start of the weld bead. As I said before, this is the part of the programming that I like to use a flashlight because the extra light allows me to get the points just a little more accurate.

The majority of welds, especially if the two materials are the same thickness, you'll want to place the tip of the wire exactly in the center of the joint, just above the metal.

The reason that you don't want the wire to be touching the metal is because when you touch the wire to the work it may bend the wire or when you begin to move towards the next point in the weld the wire may drag and bend. Bent wires will guarantee that your points in the program will ALWAYS be in the wrong place. A robot programmer must always be conscientious in making sure that the wire remains straight and extended the proper amount. If you allow either of these to fluctuate then all of the points that you programmed before correcting them will be wrong.

Once you have it at the proper location you'll need to make sure that the angle of attack is correct. The angle of attack has two main factors; the angle of push or pull and the angle between the two pieces of metal that are being joined.

Assuming that both pieces are of equal thickness the nozzle should be at an equal distance from both i.e., forty-five degrees when welding a ninety degree or perpendicular joint. If one of the pieces is thicker than the other you will need to angle towards the thicker one to get an equal burn in on the weld.

When welding steel and in fact, the majority of metals, the best welds are usually produced when you use approximately five degrees of push. There will be times when this is not possible or feasible but I try to get as close as possible to the five degree optimum.

Once you are sure that you are at the proper position then record the point. Now you have told the robot where to begin the weld so next you'll have to tell it how to weld it. This is done with an arc start command, sometimes referred to as an arc condition file.

With most robot control systems a generic arc condition page will open up with a number of zero, never program with this one. If some preprogrammed arc-start conditions have been saved into the controller's memory then you can select one of those or if none have been saved or if none of the saved files are suitable for your current project then you will need to create a new one. The common way of doing this for most robot welders is to simply type in an unused condition number. It will probably tell you that the condition doesn't exist, just click ok and you can begin creating one.

There are far too many different choices of fine tuning settings, depending on which name brand robot/controller/welder combination you are programming, the alloy you are welding or the wire you are using for me to even try to go into specifics here so I won't even try. The main two settings that you will need to worry about, especially at first, is the heat or amps and the travel speed. On the majority of robots you adjust the amps and the robot chooses the appropriate wire speed setting. That, combined with the travel speed along the weld seam is ninety-five percent of what you will ever need to adjust.

If all or most of the welds on a project are the same, or any time you have a lot of welds on any given job that will be using the exact same weld settings it is usually a good practice to create a new weld condition anyway. When you use an existing arc condition file that is already being used in other projects then you won't be able to adjust it without screwing up the other projects. If you need to adjust the settings and there are twenty welds using it on this job then you'll need to change the arc start condition numbers on all twenty welds individually. If, on the other hand, you create a new weld condition that is currently being used on no other project except this one then you won't have that problem. Then if you need to adjust those same twenty welds all you will need to do is go in and adjust the one arc start condition file and all twenty welds will be adjusted at once. You might end up with several arc start condition files stored that have exactly the same parameters but that is a small price to pay for the benefit you will receive, this is a great time, effort, and money saver. This is one of those cases where you can save yourself some work, make everyone above you happy and put a feather in your hat all at the same time, how many times does that happen?

After you have told the robot how hot to weld and how fast to travel along the seam, then you'll need to show it what direction to weld. There are only two basic directions for a robot to weld; either in a line or a circle.

A line is the shortest distance between two points. Now I realize that you already know that and I'm not trying to insult your intelligence but some people don't realize that means in any direction in the three dimensional space surrounding the robot. The robot doesn't know up, down, side to side, back and forth, diagonally, etc... The robot only knows where the two points are in a three dimensional grid and will go in an absolutely straight line between the two as long as you tell it to do so in with its highest level of accuracy.

A circle is one thing that has always been a misnomer. I say this because a robot seldom uses this to actually go in a complete circle. Most of the time you use the circle command you really want the robot to weld an arc, or rather part of a circle. The term 'Arc' was probably not used to prevent it from being confused with the arc starts and arc ends. If I had been the one naming this function I believe that I would have chosen 'Radius' instead, that would have been a much more descriptive term than circle.

Unlike the lines, which have only beginning and ending points, the circle command requires a beginning, a mid-point, and an end. This is a minimum requirement only and if you are welding a complete, perfectly round circle or an arc or partial circle that has an even radius, this is all the points that you will need. However, in a real work environment most of the arcs you will be dealing with are not quite so even or perfect. Many of them will have one or more changes in the radius and each of these changes will need extra points to keep the weld in the seam. There will also be situations when you will need extra points to keep the proper angle of attack even if you have a perfect circle. So although you have to have the required minimum number of circle commands, extra circle can commands will sometimes help but will seldom hurt the weld.

Whether you are welding in a straight line or a circle once you have shown the robot how to get to the other end of the weld the next step is to tell the robot to quit welding. This is a very important step and one that you WILL forget from time to time. If you don't catch this mistake before you start the first run you will easily recognize the problem when it keeps feeding wire and trying to weld as it moves through the air from one weld to the next. If the two welds happen to be in close quarters it can even make a mess as it arcs off on the surrounding metal, either of the fixture or the product itself. This is usually the first thing that I check for as I look back through the program before the test run. The Arc End line should be immediately after either the Line command or the last Circle command.

After finishing the weld you need to get the robot nozzle and tip away from the weld at least an inch or two, I usually like to pull straight back away from the weld. On the OTC Daihen robots this is done by choosing the tool coordinates and using the Z+ movement. All robots have this ability but it is enabled by different methods. Find out how to utilize it on your robot, this one of the most useful movements that is used in robot programming. I have used it in many different ways while writing the programs or getting the robot out of sticky situations, even pulling the nozzle back to check and/or change the contact tip and nozzle, etc…it is one very handy trick.

Another place that I use this trick quite often is once I create the point for the start of the weld I'll use this to pull straight back away a couple of inches and overwrite the previous point that I had just recorded for the approach point. This insures that the approach point is exactly where I want it, where it will dive straight in for the weld with the proper angle of attack as shown in the illustration below:

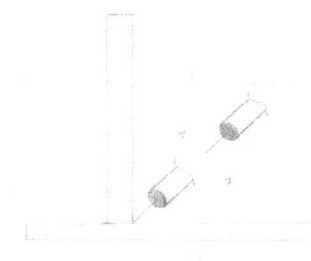 If this is the only weld to be made you'll now need to get the robot back to home base. The first point that you will need to return to is the last point before the tip cleaning position. Check to see if the robot arm can make the move safely in a single move. If it can, then just return to that line in the program and use the check go function to move the robot to that point and then return to the last line in the program, without moving the robot, and add that point to the program. If it can't safely make that move then you will have to manually jog the robot arm into a position where it can get there while recording points along the way. Once you have the robot arm back into that position you will need to return to the last line in the program, without moving the robot, and add that point to the program. If there are curtain dodging moves in this station's template then you will need to use the same method to move the robot arm and record the points in the reverse order from their place in the front of the program. Keep doing this until the robot is back into the home position. Finally add an END to the program so the robot will know that it is through with the job.

A robot programmer must always keep in mind that a robot is really very stupid. You have to tell them, step by step, every single thing to do, all the way through the program. If you leave out one thing or tell it something wrong, or even just put something in the wrong order the robot will either mess something up that you don't really want to fix or get totally confused and simply stop. That is why the next step is just about the most important step in the process of writing a program for a welding robot.

This all important step is for you to go back through the program at least twice in the edit mode. The first time will be to change the accuracy levels of the welds; all of the lines and circles need to be at the highest accuracy level possible. If you are dealing with tight movements around a fixture or other parts then you may want to adjust the accuracy level of those movements as well. The second time through the program is to look for mistakes and omissions such as missing arc starts and/or arc ends. I have found that, at least in my experience, that the best way to check for omissions is to check through the program backwards starting from the end. I have heard many theories on why this seems to work better but I believe that if you are going through it forwards you tend to read something without realizing it or mentally insert something that isn't there without noticing that you did it. However, if you are going through it backwards you are more likely to spot something wrong or omitted because you don't normally read in that direction.

Once you have adjusted the accuracy levels and made sure that you haven't missed anything then it is time for your first test run. Be sure to bring the robot out of edit mode before you try putting into playback for a run, you'll be surprised to find out just how easy it is to make that simple little mistake. The two most important differences between a test run and an actual run is that you need to turn the weld off and reduce the speed by overriding the programmed speed. I usually perform the first test run at thirty percent.

If you are using an OTC Daihen robot you can easily override the operating speed of the robot by using the Enable key. When the robot is in 'Playback' mode and you press and hold the Enable key two of the Function keys change properties and can be used to raise and lower the operating speed of the robot by percentages.

Note: You'll notice that it can also be raised to over one hundred percent, don't be tempted to use this to make the robot run the jobs quicker. Once you go over the one hundred percent mark the level indicator line turns red, it does this because you have overridden the safety margin that is built into the robot. Using it at any speed over that margin also adds a significant amount of wear and tear on the robot and will drastically decrease the lifespan of your robot. I have always suspected that this is why they have included a way to raise the speed of the robot; if you give into the temptation and use it then they will end up selling more robots. That means that it is smart for them to put it there but stupid for you to use it.

During the test runs there are several things that you need to be looking for. First and foremost is the distance the arm stays from all obstacles, if it comes close at the slower speeds it will get even closer as the speed is increased so make sure it has plenty of clearance. Any unexpected and/or unnecessary movements are another thing you need to catch and change early on. Missed and/or repeated welds are another common error that is caught in the test runs as well.

If everything looks good on the thirty percent run then bump it up to fifty percent, and then seventy before trying a full speed run. Once you do a hundred percent run and everything checks out it would be time to turn the weld back on, of course if you are practicing with plastic, cardboard boxes, or blocks of wood you'll need to skip this last step.

However, if you are practicing with real parts then it is now time to adjust the weld condition. If you happen to be using one of the newer OTC Daihen robots you can use the check weld function to check the weld in the teach mode, if not you'll just have to check it in the run mode. For the robots that have it the check weld function it is one of the handiest improvements that have ever been added by the research and development department. This enables a programmer to put on a welding shield and actually watch the robot as it makes a weld. Being able to see the arc as the weld bead is forming is by far the best way to determine what needs to be changed. Of course this applies more to the welder that has become a programmer than to the non-welding programmer.

To use this function on the OTC Daihen robots you will need to change the protection level to "Specialist". You can do this by pressing the "R" key, insert the numeric code "314", then press the enter key, type in the password "12345", and then the enter key once more. You should see a message telling you that the protection level has been changed to Specialist.

To use the Check Weld have the robot in the position of the point immediately before the weld that you want to check. Then make sure that the robot is set for continuous motion. If you are in the Easy-Teach mode you will have to press the clamp/arc key to bring the robot out of Easy-Teach and then find the Change Key function beside the F1 function key. Press (Change Key) and then turn on the Check Weld with the F10 function key.

Once it's on you should hear a beeping sound, that is the warning letting you know that when you check forward that the robot will now be welding so you need to protect your eyes. Drop your shield before pushing and holding the "Check Run" button. This will cause the robot to make the weld while you watch. If it is not welding right you don't have to let it finish the weld simply release the "Check Run" button and it will stop.

Go ahead and change the arc-start condition file, remember to only make one change at a time, and then try the weld again.

Note: After you make the changes you'll need to turn the check weld function back on again. It shuts off as a safety factor while you make any changes to the programming.

If that weld checks out and you move ahead toward the next weld be sure to manually turn it off the check weld function before checking through that next weld unless you want to check that one also. Not remembering this can cause your eyes to experience some flash burn.

Continue checking and changing until the arc-start condition has the proper settings for the weld in question. You can do as many times as necessary until you get it right. I have actually made ten or more adjustments in a single three inch weld bead.

If you have more than one type of weld on the project you will need to repeat this procedure with each of the different welds until all of the weld beads will pass their individual inspections.

At this point you have made all of the checks you can make before actually turning it loose on a run and seeing how the finished product turns out. Don't feel bad if you will still need to tweak the programming or arc-start conditions, I seldom write a program that I don't have to tweak after it starts running. In fact, most of the times I need to run the program several times to get all of the bugs out of them, that's just standard operational procedure.

I went through this just as if you were writing a program for an actual project in a solid fixture, but as I said before, I suggest that you practice programming around something lightweight and moveable so that you don't have to worry about damaging anything until you get the hang of moving the robot and programming in close quarters.

Some of the objects that I sometimes use when I am training a potential programmer are; cardboard boxes and paper cartons, plastic cups and containers, wooden blocks, soda and vegetable cans, anything really so long as it is not attached to the table or in a fixture. It's also a good idea to use objects that are light enough that they will move before the wire bends. The main key is that whatever you use must be easily moved out of the way by the robot if you happen to push a key that you didn't mean to.

When doing this type of practice the goal that you should be aiming for is to do all of the programming and all of the practice runs without moving the objects at all. Sounds easy doesn't it, take my word for it, it's not, at least not at first anyway. I have actually never had a trainee that was able to do it on the first few attempts, not even one.

I suggest that you start with a square box that measures around four to six inches on the sides. Place the box in the center of the table and have it turned so that it is square with the table. Write the program just as if it was a part to be welded, the only real difference is that you don't need to set the actual arc conditions when inputting the parameters of the weld just the travel speed. Also, you can't, of course, go beyond the test runs. Before you begin writing the program take a felt tipped marker or silver pencil and carefully trace around the outside leaving a mark all the way around. This is so that when, not if, but when you touch and move it you'll be able to place the box back in the exact same spot.

Program it for a solid weld on all four sides. Write a complete program and do the test runs up to the hundred percent run before coming back here and giving me some honest answers to a few questions.

How did you do? Rate your performance from 1 to 10. If you can honestly rate yourself higher that a seven or an eight on you first program then you did a much better job than I did on my first attempt, I wouldn't have gone any higher than a three or four.

How many times did you move the box during the programming? How many during the test runs? If either of these numbers was less than two then you did a great job.

Are you glad that you had a moveable object to program around? I wish I would have when I tried to write my first program. Remembering my first few tries at writing a program is what gave me the idea to use movable objects when I first began teaching others to program robots.

Did you write the program only using lines? If you have perfectly square corners or nearly so then the lines should connect satisfactory. However, in the real world of robot programming many of the corners that you will be dealing with will instead have a radius instead. Lines alone will usually not be sufficient in these cases and the weld will have gaps in the corners. The way out of this dilemma is to end each of the lines with two circle points going around the corner. This simple fix allows the corners to burn in and it will overlap the beginning of the first bead as well creating a good solid weld.

How many welds did you use getting around the four sides? Did you answer four? Well, that will work and in most cases there is nothing wrong with that either. If the project that you are programming is an extremely large job with many welds then it could make a difference in both the size of the program and the speed that it runs at. In most cases you can weld two sides of a square or rectangle using only one bead. This can be accomplished by making the first weld ending of course with the line command followed immediately with two circle commands to get around the corner and reposition the torch into the attitude needed to make the second weld. Repeat this for the second set of two sides and you will have welded all the way around a square or rectangle using only two welds or occasionally even one. It is sometimes possible to make all four welds around a square or rectangle without stopping but it must be located in a very small area of the station in relation to the robot which I refer to as the 'Sweet Spot'. On top of that it must also be within a few degrees of a certain rotational angle. There are several reasons for going to the trouble to program a weld like this; appearance, speed and weld strength. Any time that you remove the heat from a weld and then restart the weld it will not have the strength acquired with a continuous bead. The illustration below will give you an idea of how this can be accomplished. To actually make it all the way around using only one continuous bead requires every condition to be perfect, for that reason do not expect that to happen too regularly.

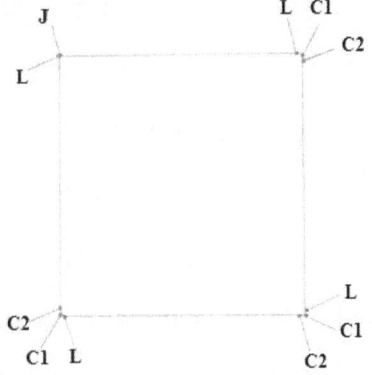

You can also do the same thing for an inside corner such as you would need welding a plate onto a length of angle iron as shown below:

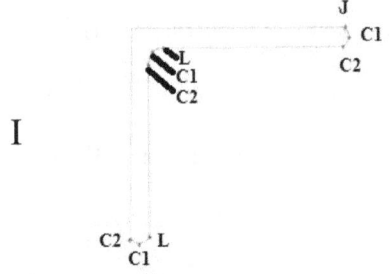

Did you have to change directions when trying to program one of the sides? I could not begin to count the times in my programming career when I had to back up and change directions because either the robot simply could not make the weld from that direction or it was bending the arm in such a way that is was getting too close to limits of travel and I figured that it would have trouble when running the program at full speed. The more experience you pile up, the less that this problem will arise but believe me, it can and does occasionally happen to everyone that is programming a robot welder.

Did you consider the natural inclination of the torch while writing the program? The torch of most robot welders are angled, you need to use that angle to help you to keep the proper angle of attack as you write the program. The direction that you choose for the robot to make the weld in is always largely determined by this angle. This means that you will need to take that angle into consideration before making that decision. With most robots the torch is angled back towards the robot while in the home base position. This means that any welds that are directly in front of the robot cannot be easily pushed away from the robot and that while you can easily reach over and make a weld on the opposite side of something you will sometimes have a problem welding the side that is on the side that is facing the robot especially if it is in close proximity to the robot.

The natural inclination of the torch will also largely determine where and how you will mount the fixture on the table. There will always be welds that would be impossible to make if the fixture is mounted in the wrong position. This is why it is so very critical that the robot programmer be involved in the design phase of the fixtures, if you don't there will be times when the programmer will handed a fixture that will simply not work. The programmers, after enough trial and failures, have gained an understanding of the very real limitations of the robot arms. There are simply places that a robot cannot reach, movements that they cannot perform; many of these are because of the inclination of the torch. Although there will be times when this natural angle will prevent you from making a weld a certain way, there will be even more times when it will work to your advantage, making it possible to reach places that you wouldn't be able to if the robot had been equipped with a straight torch instead.

Did you have to make corrections when going back through the program? Simple programs with only a few welds I can sometimes get right the first time with little or no tweaking, this was not always the case. When I first started programming it was an integral part of each job. The more complex programs that have many welds will always need more tweaking; this is just part of a programmer's job so don't feel bad when the need arises.

Did you have fun? If you didn't, then it's possible that robot programming is not for you. Programming is the part of the job you are supposed to love, need to if you are going to make it in this business. Robot programming is always a challenge, frustrating at the best of times, downright exasperating at others. I couldn't imagine sticking with it if I didn't love it. There have been times that I felt like beating my head against a wall but it is still my favorite out of the many different positions I have held over the years of my working career. In fact, the only thing that I love more than programming robots is teaching others to program them.

The rest of this chapter is going to be a series of exercises designed to get you used to programming welds in different positions and oriented in various locations in relation to the robot. It is important that you successfully complete each assignment before moving ahead to the next exercise because they are going to be increasingly difficult as you go through the chapter. If you decide to skip any of the exercises or move ahead before you have learned how to conquer the previous one, be sure and come back to it, all of these techniques are very important and you will need these skills as you move ahead in your new career as a programmer. Take my word for it, later you will look back at this experience and be glad that you followed it all the way through. I wish that I would have had an instructor that gave me this type of instruction as I was learning to program my first robot, in fact, I'm sure that the robot also wished for the same thing. I can't begin to count the number of minor crashes that I had by not starting with moveable objects and I had a couple of major ones that involved the repair of the robot itself, I'm not joking either. Be sure that you are ready before you attempt to program around an immoveable object such as real job clamped in a fixture that is bolted down solidly to the table.

Now that you have programmed around the box with it placed square with the table, rotate the box so that it is sitting in a diamond formation and program it like that. You'll find that there is a major difference when programming diagonal welds instead of straight ones, simply that there isn't a diagonal movement key to push when manipulating the robot around the cell. The easiest way to accomplish this is to first use the X or Y movement to bring the torch even with the other end of the weld or corner of the box and then use the other one to bring the torch back to the seam again. Once you have the beginning and ending points in the right place the robot will travel the seam when you check back or forward. It doesn't matter how you managed to get the torch to the other end of the seam, the robot only sees the two points and will draw a straight line between them. This method will work just as well when programming a downhill weld that is also on an angle across the table. This is because, as I mentioned earlier, the robot only knows Lines and Circles so even if you had to use movements in all three axis directions to get to the ending point of the weld, then the robot will cut straight through the middle to draw a straight line to the other point. You may also change the angle of attack from one end of a line to the other, and sometimes you may have too, but I recommend that you keep this to a minimum because you will definitely see the effects of that change in the welding bead.

As you program these angled or diagonal welds you may also want to pull the torch back away from the metal as you move the robot towards the other end of the weld to prevent the tip of the wire from dragging and therefore bending the wire, this is especially true with an uphill or a downhill, the robot will only see and remember the two points and travel in a straight line between them.

The same principle can be utilized when programming a circle; you don't have to actually attempt to move the robot in a circular motion to get to the three required points. Point in fact, the robot doesn't have circular movement keys so you are forced to get to the next points in a circle or arc by using straight movements.

Hint: Don't forget to use the rotational keys to keep the angle of attack the same, or as close as possible, as the robot moves around the circle.

The robot will only remember the memorized points as well as the angle of attack and then follow proper radius around the circle or arc.

After writing the new program with the box rotated into the diamond formation you should move the box to a different location on the station's table and write another program. The reason for this move is because the first two programs were likely written with the box directly in front of the robot, you'll find that it can be a lot different when you have to reach around to make the weld in an off centered position, you'll find that the closer that you get to the edge of the station's table the more difficult it will get to program the weld on the far side of the job. Try it at several different locations so that you can see the differences. This will also illustrate to you why it is always better to mount the fixtures directly centered with the robot when at all possible. Actually I should have said the center of the welds instead of the fixture because some fixtures will be built with the welds all to one side of the fixture.

The good news is that after you have been programming a while you'll be able to get a mental picture of the job and the welds involved in relation to the robot's possible movements and limitations when you are deciding where and in what orientation to mount the fixture on the table. However, the bad news is that this valuable skill will only be developed by the trial and failure method in which the failures will teach you much more that the successes.

After playing around with the square box in different locations and orientations then switch to a round object and practice programming around it. Do it first in the center and then again towards the ends of the table to get a feel for it. I have found that toilet tissue spools, plastic or paper drinking cups and small inverted bowls are some great round objects to use. Also, just as you did with the box, don't forget to trace around the round objects so that you can return them to their exact location after you move them. You can use anything for a practice object so long as it is light weight and easily moved by an accidental touch of the wire or torch.

Now that you have practiced with both square and round objects you are probably feeling pretty confident in your growing ability to write programs, and you should, but let's not get too overconfident either. There are probably not going to be very many projects where you are only going to be welding one object at a time to a base plate, there will be many more times when the robot will need to be making several welds each time that you push the button.

With that in mind do you remember the obstacle course that you set up in the first exercise; I want you to set it up once again only this time program the robot to make all of the different welds before returning to the home position.

How did you do on that exercise? Did you get tired of setting the objects back into place? That was a lot different wasn't it? It's usually the angle of attack that will give you the biggest problem, keeping it the same without the torch striking the object sitting next door. After you manage to get this program to running good without dislodging any of the objects I want you to compress the pattern so that all of the objects are now closer and try it again. You'll find that the closer that you place the practice objects to each other the more difficult it becomes to keep the proper angle of attack as you program the welds.

For the next exercise you will need something a bit taller, at least two feet. This is because there are going to be times when you will need to be able to get the robot arm in and out of tight places without touching anything. What I am fond of using at this point in my training is two pieces of wood that are usually referred to as finished four by fours. They are actually three and a half inches by three and a half inches and 24 inches tall. I have them cut and then sanded absolutely square on a disc sander so that they will stand perpendicular to the table.

I usually start off with them standing side by side centered on the table. The first time I will place them with twelve inches of space between them and reduce that by one inch each time until there is only eight inches left. After you can program these and the robot can do the full speed runs without dislodging them, try moving them to different locations on the table as you did the boxes. You'll find that these are even more difficult to program as you get further off centered and nearly impossible well before you get to the edges of the table. This exercise will also serve to emphasize the need to keep any fixture that requires you to thread the arm inside of tight spaces or around objects either located in the center of the table or angled so that the fixture is facing the robot's center line.

The reason that I have you to move these exercises to different locations on the station's table is to begin introducing you to the art of visualization. You will need this in every aspect of robot programming from deciding where to mount the fixture on the table to figuring out which direction to make the welds in or even what order to do the welds in.

In some cases, especially where the tables are at different distances from the robot, it will even help you to be able to determine which station that you will need to use for that particular project. For this reason I encourage you to switch stations with each of these exercises. Even if the tables are in the exact same location in relation to the robot, which is extremely rare, you will find that moving the robot around each station to be slightly different. This is due in part to the fact that each of the robot's joints is offset from the one before and after it. You will notice that the result of this is that the robot will bend in certain directions easier and reach around objects better from one side than it does from the other.

Actually visualization is one of the most important skills that you will need to develop as you advance into the trade of robot programming. It is a key element involved in the navigation of the three dimensional space of each of the robot's stations. You literally need to see, or rather visualize, the robot moving from weld to weld before you begin to actually move the robot. This is why I say that the time that you spend 'Playing Around' with the robot is so very important as you travel down the road on the way to becoming a top notch programmer. This is when and how you will get familiar with not only the movements of the robot arm but also the limits of that movement as well.

For the next step in this particular exercise you will also need to place the posts corner to corner about an inch apart, one in front of the other in line with the center line of the robot as well as in various other locations around the station, you'll find that all of the different locations will come with their own unique types of problems and difficulties. You'll find that in certain locations on the table they are easier to program but in others they are impossible or nearly so.

This is one of the hardest exercises that I set up for my students but also one of the most critical and important to learn. There are some programmers that never have to weave their robot arms into and out of tight places or even need to reach around to get to the back side of something, I actually envy these programmers. I seldom have any of these "Gravy" jobs myself. There have been several occasions during my own career that I have had to actually hand the fixtures back to the machinist because they had built a fixture that absolutely could not be programmed. This is why I always stress that the programmer needs to at least be in the loop as a fixture is being designed. The best machinist I ever worked with bugged me constantly with questions all the way through the process until he handed me the fixture, even to the point of being annoying at times. However, none of his fixtures, and he built well over a hundred for me, ever needed to be altered. All of his fixtures worked perfectly from the very start. When this man retired the company we worked for felt the experience in its proverbial pocketbook, the fixtures that his replacement turned out seemed to always need tweaking in one way or another which cost the company a lot of time and money.

After completing all of the previous exercises it is now time to switch to programming metal parts and actually turning the weld function on at last. Although you are graduating to metal you still are not ready for fixtures or even tack welded parts. You need to use metal objects that will move out of the way if the robot touches it just like the other, non-metallic, objects you have been learning to program around.

First you'll need to attach a base plate securely to the table either through the use of clamps or bolts. Cut a short piece of tubing; square or rectangular and make sure that it is perfectly flat and square on at least one end so that it will sit straight on the base. Place it in the center of the plate similar to the first position that you programmed the cardboard box in and trace around the outside in the same manner so that, as with the non-metallic practice objects, if and when you move it you'll be able to return it to the exact same location. As before make a copy of the template for that station so that you can work with the copy. The big difference this time is that instead of diving in for the first weld you will want to make some small tack welds first to hold the object in in place while you make the welds.

It makes sense that the tack weld is the first weld that you will be learning to do with the robot because tack welding is used in a lot of different ways and for a great variety of reasons. There have been many jobs that I have programmed over the years that were only tack welded.

A tack weld has every element that a regular weld has in it with the exception of movement. First you will need to establish a point a couple of inches out from the tack weld, a point showing the robot where to put the tack, a command telling it to weld, a timer telling the robot how long to weld, a command telling the robot to stop welding and then a point back a couple of inches away from the tack before moving to the next location where you will need a tack or the start of a weld bead. The only real difference between a regular weld and a tack weld is that in the place of a Circle or Line movement command you will insert a timer command telling the robot the duration of the tack weld.

There will be times, even when the part is held in place by a fixture, that you may need a large tack weld, you may even need to add a weave in to make sure that both pieces are connected good, but for now, with a free standing part you'll need to make them just as small as possible so that the part doesn't draw towards the first side before you can get to the opposite side for the next tack.

When you come in close to start the tack weld you need to be directly in the center of the joint with the end of the wire a thirty-second to a sixteenth of an inch away. Next you'll need to add an arc start. As a good rule of thumb for a small tack weld I usually have an amp setting around twenty-five to fifty percent higher than what you plan on using in a weld with the same material. Now comes the problem of determining how long to set the timer for, this is where being an experienced welder will come in handy. You will need the robot to do a hot and fast tack weld which means that a higher amp setting will result in a shorter time period that you will need on the timer. With carbon steel that is less than an eighth of an inch I'll usually start with a half second, or from an eighth to a quarter inch I usually try three quarters of a second on the first test run. After you test the tack, check to see if it is both connected good and flat. A perfect tack weld will be just large enough to hold the joint securely but without drawing and be small enough to burn through as robot welds over the top of it. That last part is very important because the customer will not want to see a bump in the weld bead but rather a smooth, continuous bead. Later in your programming career this will be equally important if the product is going to need to pass an ex-ray test or certification.

Keep in mind that appearance means as much or more to a majority of your customer base as the strength of the welds so you must always take that into consideration, especially when dealing with tack welds. Squares, rectangles, triangles, or anything else that has corners give you the perfect place to put the tacks; the corners. Even if you don't stop on a corner the robot will usually slow down as it repositions itself for the next side and will help hide the location of the tacks.

When tack welding a free-standing piece you will also need to keep the time between tacks to a minimum. One thing that will help you with this is to keep the number of open air points to a minimum. You will also need to have the first two tacks on opposite sides of the piece that you are tack welding to reduce the amount of drawing. Usually the largest chance of drawing will occur between the first two.

When you pull back from a weld you normally want to pull back a couple of inches and record a point there before moving to a mid-point and then record an approach point a couple of inches away from the start of the next weld. When you are tacking a free standing object you'll want eliminate both of those close points, leaving only the one or occasionally two midway points. You'll also want to make sure that all of the movement speeds are set on one hundred percent as well.

The minimum number of tack welds that you need to hold the part securely is usually three. The third tack weld should be halfway between the first two, if the part is not square or rectangular the third should be on the side that is the farthest from the center line of the part so that it will be more stable. Immediately after the third tack you will need to move the torch to the opposite side in the same manner and with the minimum amount of stopping points just as if you were going to put on a forth tack. Instead of the tack weld this is where you will program the beginning of the first weld.

In this case, using tack welds in rapid succession is a method that I use in training to tack weld a freestanding piece with a minimum of drawing so you will not need a fixture to hold it in place. However, no job should be run without a way of holding the part securely. There are a number of reasons that tack welds should be used in the normal operation of a robot welder. The most popular of these is to reduce the potential drawing just as with the testing exercise. All welds draw the joint to a certain degree; the hotter the weld the more it will draw. It doesn't matter how many tack welds that you put on it first or how well you have it clamped to the fixture, it will still draw some.

As an experiment I once welded two pieces of metal to a two inch thick table top before welding them together. Afterwards, I let it all cool completely before cutting it loose from the table and as soon as I cut the first piece loose from the table it sprang up. It had drawn even though it couldn't move at the time and then when I cut it loose the built-up pressure released all at once.

If a product is something that needs absolute precision and drawing cannot be allowed then you may have to use the drawing effect instead of trying to fight it by pre-flexing a joint in the same amount but in the opposite direction as the drawing. This can be a rather tedious affair which involves scrapping a couple of parts but sometimes necessary to keep the product true and accurate.

Tack welding will reduce the need for excessive clamping in the fixture. In some cases, depending on how the product is assembled, you may be able to have a fixture without any clamps. I've had quite a few fixtures over the years that had all or most of the material held in place using spring clips. Here's a good fixture tip; if you are going to make spring clips to hold tabs or small parts in place, don't buy the material to make them with. The banding material that binds most shipments of metal together is made of a high tensile strength material that is perfect to use for spring clips. All you have to do is cut them to length, bend them into shape and drill or punch a hole in one end to use for bolting them to the fixture. In the illustration below you will see two of the more popular designs.

The two main factors involved in the design of these clips is to make the section of the clip that will be holding the part in place is a bit lower than the flat surface that you use to bolt it down with and that the free end of the clip needs a slight upward turn. The first is to insure that the clip will assert enough downward pressure to hold the part in place and second is to give the clip a lifting surface so that you can slide the part underneath the clip. The upward turn also serves to give you a place that you can use to lift the front edge of the clip up slightly with a finger if needed to get the part underneath the clip.

Another popular use of tack welds is when plug welding. Small plug welds can often be done with one hot tack but the larger holes will sometimes require a series of tacks with cooling pauses to prevent burn-through or keep the weld from sagging down from the pull of gravity if the plug weld is in the vertical position.

Tack welds are also useful when you have problem areas that need extra material. If you attempt to pause and let the puddle build up as it is laying down the weld bead it will sometimes generate too much heat and either burn through or sink in which is the reverse of what you are aiming for. Putting a tack weld on prior to the start of the weld and allowing it to cool, even if it's only for a second or two, will oftentimes give you the result you are looking for. Two places that I find myself using tacks in this manner are when trying to weld something with a radius into a square corner or at the top of a downhill bead when the weld is trying to leave a dip in the very top of the bead. The latter happens most often when there is a gap in the joint.

After programming this first exercise, do as you did when programming around the cardboard boxes, move it around the table, turn it different ways and place it on different stations until you have programmed and tuned in several welds. Because you are now actually making the welds on these exercise pieces you will need to begin with new pieces each time. A suggestion, use different material thicknesses, shapes and sizes so that you will also gain experience at a variety of weld conditions. You should also notice that your guesses will begin to get closer with each attempt, the more experience that you get will help you to better estimate which arc start conditions to use for the first attempt at a weld.

Throw some different size circles into the mix as well; circles have another aspect that you'll need to learn to deal with, rotational attitude. Between the different points around a circle you'll need to keep the same degrees of push/pull to keep the weld bead nice and even all the way around. Also, you'll find that if the rotational attitude of the arm attempts to change too much between two of the points that it will greatly affect the travel speed, speeding up and/or slowing down as it twists around. This has the effect of changing the width of the weld bead, varying the amount of buildup and even the heat generation. Having more weld on one side of the joint than the other or more heat on one side can also have more than cosmetic effects, it can cause the joint to draw more or even warp the part itself, especially if it is a joint welded to a base plate.

This can prove to be one of the harder programming problems to correct, a real hair puller. This usually involves the twisting of the arm by using the rotational movement keys. Another thing that can aid you in repairing this problem is to add more circle commands around the circle or radius so that the arm movements are broken into smaller segments. The main thing with programming a circle is that the angle of attack must be kept the same as the robot moves around the circle. Another solution that I have had some success with is to jump ahead and see if I can get the robot into position with the proper angle of attack for the last point of the weld, if this is possible then sometimes you can back track and fill in the points in between.

If it's not possible to hold the same angle of attack around the circle for whatever reason, I have occasionally had to actually add another arc start condition to change the travel speed between two of the points. If you can't seem to get anything else to work you may have to go back and reprogram the weld from the opposite end. I know that involves a lot of work but all programmers have tried to program a weld from the wrong direction, it happens, it seems that none of us is immune to it. If it happens, you just have to swallow your pride and go back and start programming that weld all over again from the opposite direction.

Programming will often cause you to be very creative and imaginative as you try to fix a problem, there is seemingly no end to the ways that a robot can run into the proverbial brick wall. These are the times that you will find out if you truly have the disposition to become a robot programmer or not because this is when it either becomes fun or nerve racking. If you don't love a challenge then I'd suggest that you stay away from programming because it will constantly test your intelligence and try your patience on a regular basis.

Up until now all of the welds that you have been experimenting with have been flat down on the table, now you need to begin making welds in other positions and locations above the table level. The best way that I have found to do this is to first attach a plate to the table and then using a hand welding machine tack weld a structure together on top of that plate. The test sample can be just as simple or complex as you desire but I suggest starting out fairly simple and working your way up. I will give you a few examples of the first ones that I like to start with when training a new programmer but you will need to look at the type of products that you will be building and decide which type of welds and what materials you need to incorporate into the exercise samples.

I usually like to start out with a box; four sides and a cap using real small tack welds that wouldn't interfere with the weld or programming. Have the pieces cut so that you can connect the inside corners and leave an outside corner for a good fillet weld. This simple structure will give you four downhill welds and four horizontal welds that are above the table as well as four that are on the base. I usually assemble them in pairs with one straight with the table and the other diagonal so that one of the corners is facing the centerline of the robot. Below you'll find an illustration of the type of box that I use for these exercises. Generally I will build two or three pairs of these boxes from the different gauge materials that the person that I am training will have to work with. If you are going to be working with different materials then I suggest that you build at least two using each of those materials for a more rounded training experience.

Downhill welds are some of the hardest to get right so I usually use more of them in my test structures than I do the other welding positions. If your company doesn't use downhill welds in their product line then you should count yourself lucky but I would still incorporate at least some of these difficult welds into your practice to gain experience in case you may need to know how to handle them at a later date.

I always tack a piece of round tube to the side of a vertical riser to create a circular downhill. This exercise can be made even more difficult by having the round tube short enough so that the inside can also be welded or by coping or saddling the round tube onto the side of round riser. For an added dimension in an exercise have the risers, round or square, coming up on a forty-five degree angle and then branch off of the side of it.

When you are designing test structures always strive to push up against the limits, build some that you don't think you'll be able to complete, you may just surprise yourself. You never know what type of situation you may run into on the next project or what skills will come in handy.

I once had a project, not an exercise but an actual job with runs numbering into the thousands, where I had to write a program for the robot to weld a square cut oval tube onto the side of a round tube. This was one of the hardest welds that I ever had to program in all of the years that I have been doing this. The top and bottom were connected with no gap at all but as you traveled down the sides the gap grew steadily wider until you reached the center and then progressively smaller on the way to the bottom. By the time that I finally finished tweaking the weld I had incorporated several changes in the amps and travel speed as the weld progressed from the top to the bottom on each side. I also had to begin the weld without a weave, add one in the first quarter and then widening the pattern as it encountered the wider gap in the center. During the third quarter of each side I had to reverse this procedure and narrow the pattern until I ended up without a weave at the bottom. To top it off, all of this was just to get the first side welded but it generated so much heat that all of the settings had to be different on the second side to deal with the pre-heated material. The worst part of all was the knowledge that all this could have been avoided by simply saddling the second tube into the first for a good fit without a gap, that's just poor engineering.

As a robot programmer situations like this are not uncommon and I'm afraid that I can't tell you an easy way to deal with it. Some engineers are simply too smart for their own good, and yours as well. In the case that I described above I finally talked the powers that be into coping the ends of some test tubes so I could show them just how much faster, cleaner and stronger the product would become once the tubes were saddled into the round hubs. It cut the weld time by just over seventy-five percent, strengthened the weld by fifty percent and the welds needed none of the cleaning that was being done to each and every piece before it went out the door.

After demonstrating all of these benefits that would have been a direct result of such a small change, nothing was ever done about it. The company is still building that exact same product, by the thousands, using the same flawed engineering as they always have and quite probably always will. Not even the lure of building a much better product while actually making more money with less effort was enough to get them to admit that they had engineered the job wrong and correct their mistake. In other words, sometimes it's not the robot, the jobs or the fixtures that will make you feel as though you are beating your head against the wall but most of the time it will be.

What I usually do at this part of the training is to take a good look at all of the products that are going out the door and begin tack welding different structures together that incorporates all of the likely welding positions that will be needed on them. I usually build each test structure in increasingly difficult configurations with the last few looking like some sort of leafless trees. Be sure to include a lot of welds that are real close together and some with increasingly tighter spaces to wiggle the robot arm in and out of. As I said before, you just may surprise yourself when you discover just which welds that you are actually able to program or how few are truly impossible. Always endeavor to be more stubborn than the robot and you will likely prevail.

After you have practiced with all of the weld positions that you feel that you are likely to encounter it is time to take the final step and begin programming the actual jobs. One thing that I can guarantee you is that no matter how thorough and comprehensive your training has been or how many different types of welds that you programmed during your practice sessions, you will find many more that you didn't even imagine once you begin programming the actual jobs. There has seldom been a whole month that passed since I began programming robots when I didn't have to attempt a bit of programming in a way that I have never even imagined it being done before just to get the job done.

Getting around the fixtures and the clamps as you maneuver from weld to weld will add even more new dimensions to your programming experience and skill level. It will probably be a little nerve racking, especially at first, but you just need to be a little more stubborn than the robot, occasionally it will even take more than a little, they can be very stubborn themselves.

Hint: Use the rotational keys to flex the joints around the obstacles, it's usually the third and fourth joints that will give you the most trouble in the tight spots and the rotational keys will cause these joints to twist out of the way without changing the angle of attack.

A career in robot programming is a constant learning experience, every time that you think that you have seen it all, that's when you'll be handed a job that will prove you wrong. You are now on the path to a learning curve that never ends. During the rest of this book I'll try to give you some pointers and tips to help you along but for the most part you are going to be your own teacher from this point on.

Good luck.

Tweaking to Improve Performance

Now that you are writing and running real programs for actual jobs and gaining experience it's time to begin fine tuning your skills. Once you get a program working with good welds you will be tempted to leave it alone, don't give in to this temptation. There's an old adage that says "If it ain't broke, don't fix it.", this does not apply to robot programming. There is no such thing as a program that cannot be improved on; there are always adjustments that can be made, this is a process that I call 'Tweaking'. Tweaking consists of nothing more than small, fine tuning corrections that will result in a program running faster, smoother and with better welds, producing a better product.

Satisfaction is the difference between a run-of-the-mill programmer and a top notch programmer. The best programmers are NEVER satisfied with their programs but are persistent in the quest to find ways to improve on their work. Sure, there's a certain amount of pride that goes along with the completion of any program, especially with the more difficult projects, just don't let it get in your way or allow yourself to become satisfied.

While it's true that you would never want to take a chance on messing up a program that is producing acceptable products, you don't have to be contented with it either. The safe way to try and improve a program is to make a copy of the program in question and experiment with the copy while leaving the original alone. If you do manage to get the copy working better than the original then all you have to do, with most robotic controllers, is to save it with the program number of the original and it will automatically replace it.

There are several ways to improve a program but is usually comes down to just three main options; how fast the program runs, the cosmetic appearance of the finished products or the strength of the welds.

One way you can speed up a program is to change the accuracy rating back to the minimum settings in the parts of the program where there is no chance of the robot colliding with the fixture or part. The robot will always choose the fastest path for itself if you allow it to.

Another way you can increase the speed that the robot moves through the program is to remove any unnecessary movements. This is done by minimizing the number of points between any two needed positions. Sometimes, as you try to reposition the arm for the next weld you may record five or six points in between the two, that's natural, but after you get the arm into position you should go back and take out any unnecessary points. Many of these times that move could easily be accomplished by using only three or possibly even two of them.

Removing these extra points and allowing the robot to consolidate the movement can make a huge difference in the time it takes to run the program, especially if there are quite a few places that have some of these unnecessary points throughout the program.

You must be very careful to make sure that they are in fact, un-needed points, before you delete them. One of the best ways that I have found to do this is by skipping over the points that you suspect are unnecessary in the program and then force the robot to move in between the two points that you intend to leave in the program before actually deleting anything. Be sure to do this in both directions, forward and back, before you actually delete those points. So long as the arm remains in the clear then it should be safe to remove the points in question.

Never assume that any point is not needed, check it out thoroughly to be sure. Even after you delete the point(s) always jog the robot back and forth through that part of the program a couple of times to make doubly sure that it still works as expected before moving on, this way you can always use the 'Redo' function to replace the deleted points. This is one of those cases where it's always better to be safe than sorry.

Even though fine tuning the open air movements in a program can make a real difference in the speed that a program runs, it is the actual welds that are always the slowest part of the program. The good news is that because of a robot's inherent accuracy it can be made to weld much hotter and faster than a human can possibly do by hand.

This can be accomplished by bumping up the amps and then finding a travel speed that will accommodate the extra heat without any adverse effects. You can keep repeating this procedure until you find the highest amps that can be used without burning through the material, undercutting the weld or drawing the part beyond acceptable limits. A hotter, faster weld usually produces a smoother, better looking bead as well as achieving more penetration and a stronger weld so you will be helping the cosmetic appearance as well as the overall quality while reducing the time it takes to complete the weld. By tweaking the welds in this manner you are managing to improve all three faucets of the weld; speed, strength and appearance.

When you first begin to tweak a weld or an Arc Start condition you'll probably do so in increments of five or more amps at a time but as you see that you are getting closer to the limits for that weld you'll likely reduce those changes a single amp or less. One of the best things about robot programming is the ability to use such small increments of change when you are fine tuning a weld, even to fractions of an amp.

I would never suggest that the quality of a weld be ignored but to a non-welder, which a large percentage of the customers are, having neat, clean uniform weld beads will actually mean more than the shear strength of the welds. One of the best ways to improve the appearance of the robot's welds is to reduce or eliminate the weld spatter.

The most common cause of weld spatter is an inaccurate travel speed; too fast or too slow, either one can cause it. Another thing that can cause excessive weld spatter is the angle of attack, more than ten degrees of push tends to throw the spatter up out of the puddle. Other factors that can cause excessive spatter buildup are having the weld nozzle too far away from the work, welding towards a vertical surface or into a corner, pulling instead of pushing the arc, etc…occasionally one or more of these factors may be beyond your control. There are going to be times and situations when you will simply have to minimize the spatter all that you can and then use a good anti-spatter spray and clean the joint up afterwards.

The worst cases of unpreventable spatter buildup that I have found is with pre-sanded material, especially if it comes from a machine that has somewhat ironically been nicknamed a "Time Saver". Steel that has been sanded or brushed prior to the arc welding process is a virtual spatter magnet and should be avoided if at all possible.

Yet another way of improving the appearance of the welds is by improving on the uniformity of the weld beads. If you have several similar welds but you have the robot programmed to weld them differently, the product is just not going to look as good to the customer even if it's something as simple as one of them being welded in the reversed direction.

Although the strength of the welds should never be overlooked or ignored, the beauty of the welds and of the finished product will always carry at least as much weight if not more with some of the customers who, as we all know, must be kept happy no matter what.

On the newer robot welders you have many more weld settings that can be adjusted than they did on the older models. At one time or another I have tried changing each one of them. To be honest, most of them made little or no difference in the weld, not enough that I found to be worth the effort anyway. Most welds can be fine-tuned to produce a beautiful, clean, high strength weld by modifying only the amps, travel speed, wire extension length and the angle of attack.

The angle of attack actually consists of several different factors all tied together; the angle of push/pull in relation to the weld, the angle of the nozzle in relation to the work from the perspective of all three of the axis attitudes and even to some extent, the wire extension length. However, it has always helped me to separate the wire extension length from the angle of attack and so from this point forward when I refer to the angle of attack in this book I will only be referring to the various angles of the nozzle.

With most metals the robot will produce the best welds when it is pushing the bead at a pitch of five degrees. I realize that this is sometimes not feasible or even possible but the closer you get to this ideal angle of pitch the better off you will be.

The wire extension length should remain the same regardless of the position of the weld or the torch. The optimum for welds below two hundred and fifty amps would be one half to five-eighths of an inch of wire past the end of the contact tip. For the hotter welds with settings over two hundred and fifty amps this should be increased by approximately one eighth of an inch. As with the angle of push/pull, this is a best case scenario in the best of possible worlds not the real world of robot programming. There are undoubtedly going to be occasions when you will be forced to extend the wire further to reach a place that the nozzle won't allow, in these cases just do the best that you can.

The angle of the nozzle in relation to the work is the one of these factors that WILL sometimes need to change because of the position of the weld, material thickness and other factors.

One of the most important things to remember is to only change one element at a time. The reason that this rule is so important is that this is the only way for you to know for sure if the change gave you the desired results. If you change more than one thing at a time then there will be no way to determine what effect each change had on the weld. If, for instance, you observed no change at all then one of the changes could have been positive while the other one was negative resulting in one change cancelling the other one out. Even if the overall performance was improved you would have no way of determining if one of the changes might have made even more of an improvement on its own. It is always best to only change one element at a time and then either keep the change or reverse it before moving on to the next.

You also have the option of adding a weave to the weld. I usually leave this as a last resort. As long as the joints fit well and there is no significant gap to fill up then I can usually do better without a weave. I have always found that weaves are hard to control. There are too many elements that can be adjusted and unlike the weld conditions, every little factor seems to make a big difference. To make matters worse the weaves never seem to do exactly what you expect them to do. I usually use a weave only when I am forced to or if nothing else that I've tried seems to work.

If you do find that you must use a weave then you need to create a weave condition file. This will enable you to use that same weave pattern in any subsequent program by only inputting the file number of that weave.

There are three main characteristics that control the major aspects of a weave; the type, the frequency, and the width. Most of the better robots today have three types of weave patterns to choose from; Linear, Sine and Circles as illustrated in the diagrams below.

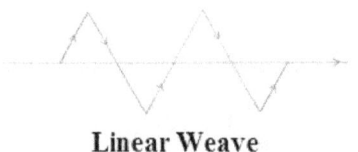

Linear Weave

Sine Weave

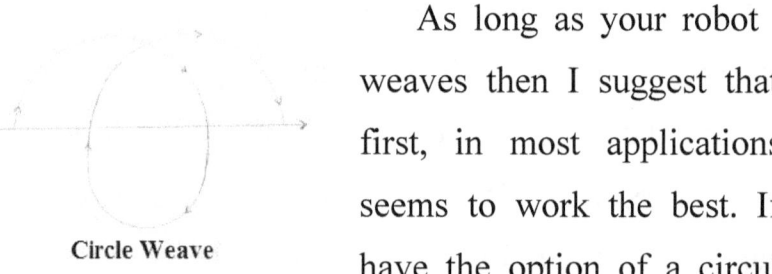

Circle Weave

As long as your robot has circular weaves then I suggest that you try it first, in most applications this type seems to work the best. If you don't have the option of a circular weave I have found little practical difference between the linear and sine weaves but the linear always seems to do a slightly better job. I have tried the sine weave on numerous occasions but I can't remember an instance where I left it in any of them.

Frequency is the amount of weaving motions per second. Most robots will give you a wide range to choose from but I have found that if you keep it at or near the twenty-five percent level you want need to move it up or down too far to find the frequency that you need.

Actually, I can't remember ever using over six and four and a half to five is all you should need for most applications. If you ever set it even close to the fifty or sixty percent level you will see and actually feel the robot trying to shake apart. If the frequency is set over seven you can feel the vibration through a concrete floor from across the room. This brings me to the other reason that I usually leave the use of weaves as a last resort; the wear and tear on the robot is tremendous anytime you use a weave. Weaves can reduce the effective lifespan of a robot welder by a significant degree.

The width of the weave is usually referred to as a right and left radius. The left and right designations are seldom where you will expect them to be. The orientation will depend on the direction and position of the weld combined with the angular position of the robot's joints within the three dimensional space that the robot operates in. Taking all of that into consideration, Left and Right can be in virtually any direction at all; right and left can be reversed, right may be up while left is down or vice-versa. Most of the robot operator's handbooks that I've ever saw claim that right and left should be oriented according to the direction of travel of the torch as shown in the illustration below:

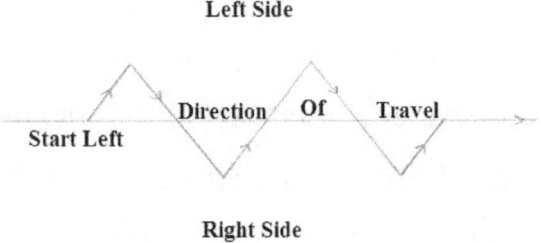

Don't believe it, I've saw times when it acted exactly opposite of that.

The only way to be sure is set it off centered and either check through the weave or simply try it out with the weld off and see if it goes in the direction that you expected it to, don't depend on it though, and check it out just to be sure. So far, every OTC Daihen robot that I have programmed acted correctly but I'll continue to test each one to be sure. Mercifully, most of the time that I have ever been forced to use weaves I needed them centered, equal on the right and left side of the weld bead so the problem didn't arise.

The main types of welds where you will need an off centered weave are situations where you are joining a thicker metal to a thinner one or on a horizontal weld bead where you are using it to fight the effects of gravity. It's usually when you need to be on one side of the weld longer than the other.

Another adjustment factor that has the same orientation problem is (Start Right) or (Start Left). It usually doesn't matter which side of the weld bead that the weave begins on but if you find the need to control it you won't know which side is right or left until you try it. If you have an option in the weave settings to make the robot weave when stopped this can be very helpful when tack welding a joint with a gap or pausing for a moment to build up a puddle before continuing the weld.

Most robots will give you even more options when setting up a weave but I recommend not changing them from their default settings or trying them out very carefully when there isn't anything close enough to slam into. Two of these "other" settings are (weave angle or torch angle), and both can easily cause the robot to crash or if you are real lucky it may just kick the motors off and throw up an error message.

Circle rate is another option that a lot of robots offer as a weave adjustment that I find to be virtually useless; it only seems to take away from the weave condition that you just took the trouble to create. I always leave it set at its default of one hundred percent.

The majority of the time that I am forced to use a weave in a weld it is because of either poorly cut pieces or parts that have been machined wrong, in other words, I'm using a weave to fill in gaps that shouldn't have been there in the first place. There is however other uses where a weave can help you accomplish the goal you are aiming for. A weave pattern can be used to widen a weld bead when slowing the travel speed or increasing the amps are not viable options. You can also use a weave to achieve better penetration on an inside corner weld or breaking the edges on an outside corner. Weaving is sometimes the best way to connect different gauge metals regardless of the position or attitude of the weld.

One of the most popular reasons for trying a weave has always been, at least for me, when nothing else seems to work while I am tweaking a weld in. The bottom line when I am dealing with the question of weaving or not is to use a weave when I must to achieve the desired results, but as long as I can get by without one, then that's going to be my choice.

Creating and fine tuning the welds in a program is usually the hardest part of the programming experience as well as the most time consuming. Over time, as you gain experience, it will give you a good head start in your estimations but it will always be a guessing game. Oh, there will be times when you'll be working with the exact same material and the welds will also be the exact same type and you will know exactly which arc start settings to use, those do tend to be the exception and not the rule though. Most of the time I take my best guess and then fine tune the weld until I get it right, occasionally I'm still tweaking the weld after I begin running the program.

There are so many factors other than the weld settings that must be taken into account; weave or no weave, type of weave, angle of attack, angle of push/pull, wire extension length, the fit of the joint, is the joint beveled or not, etc… Slight changes to any of the factors, many of which may be out of your control, can make a major difference in the quality and appearance of the weld. Of all the different types and positions of welds to program I have consistently found that the downhill beads have nearly always proved to be the most tedious and problematic. Gravity, which has little or no effect on the base weld, is your greatest enemy on a downhill weld.

If you have to wrap across the top or begin a weld on the top edge you may find that you will end up with a dip or sunk in spot in that top edge. This occurs most often when you have a gap in the joint. If the bead is otherwise a good weld I have found that the best fix is to put a small tack on the problem spot and allow it to cool slightly before beginning the actual weld. This is accomplished through the use of timer, putting a short pause into the program immediately after the tack weld which will give it time to cool, a second or less is usually quite sufficient. There have been occasions, when dealing with larger gaps, where I had to repeat this procedure two or three times to get enough weld built up so that it will not burn through. You may even find that you will have to use a lower amp setting in the tack weld for the same reason.

Another helpful trick that I sometimes use in a downhill weld with a good fit is to start the weld but add a momentary pause to allow a puddle to form before I let it begin the downhill motion. Ideally, what you want is to have the tip of the wire remain in the center of the puddle all the way from the top to the bottom of the weld. If the puddle begins to outpace the wire then it's going to start dripping balls of slag as it progresses as well as undercutting the material at the edges of the weld. This creates a mess on the part and/or fixture while leaving behind a weakened weld. If, on the other hand, the wire begins to outpace the puddle there will be very little penetration, leaving behind a thin bead without any real strength, it may even have visible gaps in the weld that would look bad to even a non-welder. To prevent either of these things from happening, a downhill weld must have the amps and the travel speed matched perfectly. When you are fine tuning a downhill weld and begin to get close you will find that the difference of one amp in the heat setting or a tenth of an inch in the travel speed will make a noticeable difference. This sensitivity is what can make programming downhill welds such a head ache.

By the book, all steel MIG welding should have a five degree push regardless of the direction you are welding in. However, I have found that to get a downhill weld just right I sometimes reduce that to two or three degrees. I also find that keeping the wire extension length a bit shorter than normal can also be beneficial, of course this may mean that the robot will burn up a few more nozzles and contact tips but that IS why they are referred to as consumables.

Gravity is also the main culprit that you must combat when you are programming horizontal welds as well. Just as with the downhill welds, the Amps and Travel speed need to be near perfect. You also cannot run a heavy bead in the horizontal position; if you do happen to need a larger weld then you may need to do so by using multiple passes. This is usually associated with chamfered joints where the edges have been beveled to allow a more complete penetration. In these cases I usually begin with a small root pass followed by a more substantial fillet pass. If you find that you must use multiple fillet passes, which is common with your thicker metals, always run a lower pass first followed by one above it, with the two passes overlapping in the middle. This is done so that the lower pass, which has cooled slightly, can add support to the upper one, and to insure that the beads adhere to the beveled edges as well as each other.

A circular weave can also be utilized to help in these situations but you'll need to experiment with the direction of travel and the offset of the weave to achieve the desired results. As a general rule of thumb you will need the weave pattern to push upwards on the forward edge of the circle and coming down on the back edge as illustrated below. Also shown below is the proper orientation but don't trust it, I have saw it completely reversed from time to time, test it each time to be sure.

Left Side

Right Side

You'll also want to offset the pattern so that it will be slightly above the centerline of the bead to help combat the gravitational force pulling down on the liquid metal.

The base welds, any weld that is either performed on the base plate or a flat weld above the base, are by far the easiest to adjust. This is because in this position gravity is actually working in your favor in all ways except on, spatter. With any of the other positions the spatter tends to fall away but with a base or flat weld it simply has nowhere else to go. Once you have adjusted the travel speed and angle of attack to reduce the production of weld spatter as much as possible a good anti-spatter spray can keep the rest from adhering too securely and make cleanup much easier.

Regardless of the type or position of the weld you are fine tuning or tweaking, the first two things that you should always check is the wire extension length and the angle of attack. This is because even if you have found the absolute perfect combination of amps and travel speed you would never know it if either of these elements was off appreciatively and so you may continue in vain to find the correct combination. Separately or in combination these two factors can distort a weld to such a degree that it would render any other adjustments a useless waste of time and energy.

Honing your Programming Skills

The welds may be the slowest part of a running programming but moving around the three dimensional space inside the robot cell is usually where the most of the time is consumed when writing the programs. This is especially true when you are attempting to maneuver through the tight spaces in and around the fixture and/or component parts.

I have found a neat little trick to help you see where you need to place the robot torch and how to get it there. Take a spare nozzle and hold it in position by hand, this will aid you in visualization. If you have already saw where you need to place the nozzle and the orientation you will need to aim for it will make a huge difference when you begin to maneuver the robot arm into position for the weld.

You can also use this same trick to help you determine if a weld is even possible. There are times when the machinist that is building the fixture will not leave enough room between the fixture, parts and the clamps to get the torch in position, with the correct angle of attack, to make the weld successfully. There have been other times when there wasn't even enough room for the nozzle alone. Just by manually maneuvering the nozzle in and out of the available space will help you visualize or mentally see the problem as well as helping you to determine what will need to be changed to eliminate the problem.

Visualization is the key to successful programming; you need to see a move visually in your mind, to actually create the move in your mind, before you try to recreate it using the robot. As you gain programming experience you will discover that you have been building, without attempting to do so, a three dimensional map of the robot's spatial area in your mind. This is a critical step in the development of your programming skills. This mental map will enable you to get the robot arm into the required position and orientation with only a minimum amount of effort.

Actually, you will end up thinking in four dimensions; the X, Y and Z axes plus the added dimension that is the orientation of the robot arm as the fourth. Once this occurs you will then be in a small but very elite group of people. Most humans go through their entire life thinking in only two dimensions; some even expand their mental capacities to the point that they can visualize in three dimensions, but only a handful of individuals will ever develop the capability to think in four dimensions. Some examples of this elite group are; pilots, astronauts, submarine navigators, mathematicians, robot programmers, etc…not a bad crowd to be in.

One of the most common uses for this four dimensional thinking when programming a robot is in preparing for your next move. This is, in many ways, very similar to the use of 'English' on a cue ball when shooting a game of pool or billiards. As you make your shot you will be attempting to leave the cue ball in a good position for the next shot.

The exact same type of strategy is used in robot programming, except that this particular game of pool has a three dimensional table. As you leave the last completed weld and head for the next one you will want to go ahead and begin reorienting the robot arm where you can, and as much as possible have the nozzle in a position to dive down towards the starting point of the next weld.

There is a neat little trick that I often use to ensure that the point immediately preceding the starting point of a weld is in the best possible position to go in for the weld. What I will do is first record a point in an approximate position above the beginning of the weld. Then I'll move in and program the beginning point for the weld. When a programmer creates this point they invariably adjust the orientation of the torch to the proper angle of attack. Afterwards, I'll pull the nozzle straight back away from the joint about two inches keeping the torch in the same angle of attack. Then I just move back through the program and overwrite the position that I recorded for the point immediately before the start of the weld. If you happen to be programming an OTC Dahien robot then just use the Z+ key with the robot coordinates set on the tool movement setting to pull back with. Doing it this way guarantees that the torch will be in the proper angle of attack before moving to the starting position. All robots have the ability to move in this manner but it will not always be labeled as the 'Tool' setting. This simple pulling away from the work movement is something that you will find yourself using for a great many different reasons as you program and operate a robot.

This type of movement is so helpful that I always make a point of showing the operators that I am training how to do this so that they can use it for checking, cleaning and replacing the contact tips and nozzles when recovering from an arc start failure.

Never create too many points in a program before checking back to insure that the robot will be able to operate both safely and smoothly through the program. Checking back doesn't take much time and it will save you many headaches as well as lost time that will occur if you have to go back later and fix a problem. The further you go before checking back, the more trouble that you are likely to run into and the more difficulties you will experience as you try to correct it. It only takes a few seconds to check back but precious minutes to fix a problem later, so check back regularly and often.

If you attempt to check back from the point that you just programmed to the previous point and you notice that the robot is about to collide with something there is a good chance that forgot to record a point. Release the check back, hopefully before the collision, and use the X, Y and Z movements to get the robot clear before trying to check back again.

Note: If you get past the object that the robot was about to collide with manually then the check back operation will usually be able to complete the move safely.

Repeat this procedure as many times as necessary to get back to the last point that you actually did record. From this point you will either have to move that point or create at least one new point in between the other two so that the robot can move safely without a collision. Most of the time this can be accomplished with one point that is even with but to the side of the object in question unless that would put the arm in danger of colliding with something else. Sometimes, when you are attempting to get the arm in and out of tight spaces there will be a very narrow path that the robot will have to follow. In cases like this you may find that when you try to prevent a possible collision you could very well be sending the arm towards a collision with something else.

When dealing with extra tight spaces you may find that you will need to add a few additional points to help the robot maintain a minimum safe clearance.

There may also be occasions when you will discover that it is not the location of the points but the orientation of the arm that is causing the problem as the arm maneuvers between the fixture and/or component parts. I have found that most of the time this problem arises it is with the fourth joint. I'll notice that the writing on that section of the arm is upside down, that is usually a bad sign. This means that you are real close to the natural limitations that are built into the robot and once you are switched into playback mode there will be no predicting how the robot is going to move.

If you notice that the arm is in such a position the best thing for you to do is to back up a few moves until that joint is back in its normal orientation and then try once more, this time finding a way to keep the arm from twisting like that again. Many times you will find this happening when you are trying to get from the beginning point to the end of the weld, this can prove to be very hard to correct, sometimes impossible. Many times I have had to go back and start the weld from the opposite end. Things like this can be a little frustrating at times but that's the world of robot programming and like it or not it is not an uncommon occurrence, this is just part of the game.

If the arm is swinging into a new orientation as it is moving between the points it can sometimes go in directions or bend in ways that you did not even expect it to. I have usually found that it is best not to allow these kinds of movements to take place except with open air movements that are 'In the Clear' and even then I'll try to hold them to a minimum. If there is the slightest chance of a collision I will force it move through the points without changing the orientation of the arm, even if it means going back and rewriting that entire section of the programming code. Only after the robot is completely clear of all obstructions will I let it swing into a new orientation. Never give a robot a chance to collide with anything; that can be very expensive because robot parts do not come cheap.

After any major changes in a program you should treat is as if it was a new program again. This means doing test runs with the weld function turned off and at a reduced speed. You may not need to drop it all the way down to thirty percent but I never do the first run higher than fifty percent of the programmed speed, but you'll need to use your own judgment. A crash at fifty percent or lower won't usually damage the robot but at seventy percent or higher that risk not only exists, but the chance is rather high. As far as I'm concerned, if there is any chance of crashing the robot I'd rather err on the side of caution.

The points that usually need to be fixed usually involve the open air movements that travel close to the fixture, part, doors or the walls of the cell. The door is the one that seems to give me the most trouble of all. I think that this is because the door is not there when I am writing the program. This is especially true of the cells that have doors that are raised and lowered by a single cylinder located invariably in the center of the door. This is because cylinders are protected by a steel box that protrudes close to two inches inside the door.

If the arm comes close to or actually hits something during a test run, back it up and run it again as you keep your finger on the stop button. This time try to stop the robot just before it reaches the point of impact or its closest approach. Switch the robot into teach mode and jog the robot through the program stopping exactly at the trouble spot. If this happens to be close to an existing point then consider yourself lucky and just move the robot slightly and overwrite the point. Usually the problem area falls in between two points. Sometimes you can still just move one point or the other but occasionally you'll have to move both or create a new one in the middle.

You'll also find that there are times as I mentioned in the last section that the problem will not be in the location of the points but the orientation of the arm. This usually happens when I attempt to reorient the arm too much in a single move. If this is the case you can usually correct it by either relocating the arm until it is 'In the Clear' to perform the reorientation move or by creating extra points and then dividing the problem movements into smaller segments and then spreading them between the points.

You should also double check to make sure that the accuracy level is at its highest level. This gives you more control over the path of the arm instead of allowing the robot to choose its own path. The illustration below shows the differences between the different accuracy 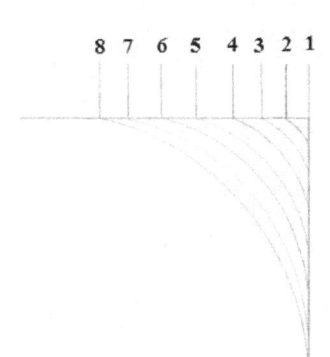 levels. The lower level numbers actually have the highest accuracy rating that the robot can be held to. At a level of one the robot is made to travel precisely through the points as you recorded them. One exception to this rule is when the robot is attempting to make a reorientation move. This happens if you try to force the robot to get too close to one of its limits. On the other hand, when set on a level of eight, the robot will pass through a wide arc on the inside of the programmed point. On level eight the robot runs the fastest but you have the least control of the robot's actual movement. All of the other levels pass through at various degrees of accuracy, each level slowing the robot down but giving you more control of the robot's movement through the program.

Another good rule of thumb is that if there is no danger of collision; let the robot decide how to get from point A to B, it will speed up and smooth out the run. However, as the quarters get closer and tighter you'll need to take back more and more of the control. I'd much rather slow the robot down and let the program take a little longer to run than to take a chance on crashing and maybe damaging one of the expensive robot parts in the process.

If the problem is with the middle sections of the arm getting too close to something then you will probably need to make use of the rotation movements to reconfigure or reform the arm so that it is back in the clear again. By using the rotation movements you can generally cause the arm to bend in ways that will enable you to get the arm in the clear but it's usually accomplished through trial and error. Try only one movement at a time, if it doesn't move in the way that you need it to don't try to reverse the movement and get it back close to where it was. Use the check back and then check forward to put the robot back into the exact same place before trying again. This will insure that the robot will be moving away from the same exact same spot with each try.

What you are trying to find is the movement that will bend the robot arm in the way that you desire without changing the angle of attack. If the robot is close to tripping its limit switches it may take a combination of several movements to get the results that you need.

Don't fret if it takes you a while to find the right combination of moves to get the program running the way that it should and don't let anyone try to rush you through the process either. If someone tries to hurry you along then remind them that if you spend a little extra time now to get the program just right, then the robot will sit there and churn out a hundred thousand parts or more with very little or no help from either you or them.

Rushing and robots don't mix very well. If the robot is running a program then it can't be rushed anyway. By using multiple stations you will be able reload a fixture on one station while the robot is running a program on the next station. So long as both jobs are fairly equal then the robot will be going from station to station without a pause, this is as fast as it can possibly run and the best that you will be able to accomplish.

If you rush through the programming phase then the program will not be as efficient and will actually run slower, this will defeat the purpose in rushing in the first place. Never skip the test runs to speed up the process either, and if you are told to do so just remind them that if something is wrong and causes the robot to crash and break something, then they'll then be dealing with downtime. If you take your time and do your job right the robot will take care of the hurrying.

I've explained how that visualization is needed in every step of programming a robot from the writing of the program all the way through the tweaking and even to the trouble shooting but it's also needed before the programming as well. When you are mounting a fixture on the robot for the first time you'll need to visualize how the robot will bend and navigate through the fixture to make the welds.

This is a lot harder when you first begin your programming career mainly because of your lack of experience. The good news is that as you gain this much needed experience, positioning the fixture will become a lot easier. The bad news is that this experience is usually hard won and acquired mainly through failure.

If you are having trouble figuring out where on the table to mount the fixture or which way to turn it, take your best guess and place the fixture in that position and then attempt to maneuver the robot to the weld you think will be the hardest to reach before you actually attach the fixture to the table. Although this tip can help prevent you from mounting the fixture in a position that won't work, it doesn't come with a guarantee either. I have, on occasion, after doing this, still had to end up moving a fixture because I wasn't able to complete the programming. Sometimes, regardless of how careful you are, the simplest things are occasionally overlooked.

Because of the possibility of needing to move the fixture to a different location on the table I never stencil the grid location onto the fixture until after I complete the programming. I will write the location into a remark line of the program because that is easy to change. If you do happen to start programming a fixture and get much of it programmed before discovering the need to move it, with some of the newer robots you won't need to start over. These robots will allow you to simply shift the program around the X, Y, and Z pattern of the station to match the new location. If you are lucky enough to be programming one of these robots you can then just tweak the relocated program and finish writing it.

I have also been handed fixtures that could not be programmed or run with the fixture mounted straight on the table. If this happens you can have one of the mounting holes elongated into a slot so that you can use one of the other holes on the table's grid to mount it with when you get it turned. Sadly, if you are already into the programming when you find this out, then the location shift won't help you save your work and you will have to start from scratch.

There is another situation where the location shift won't help you; that is if you find that you are going to have to reverse the fixture one hundred-eighty degrees. There have actually been several occasions when that was the only option which allowed me to complete all of the welds. A couple of those times I was more than ninety percent complete with the programming when I realized this and had to start from scratch. This is yet another case where Murphy's Law kicked in, "If anything can go wrong, it certainly will, and it will always happen when and where you least expect it to."

There will be times when you discover that the amp setting or travel speed that you use at the start of a weld is not the same as you need during middle of the weld or at the end of it. This can be accomplished by the nesting of Arc Starts and Arc Ends.

You always begin a weld with an Arc Start command but if you need to change the conditions of that weld all you have to do is jog the robot through the weld to the position where the change needs to take place and insert either a line or a circle into the program, whichever is appropriate, and then follow it with a new Arc Start command that has the different weld parameters in place. You can have as many Arc Start commands in a single weld as you want and/or need, I have actually had quite a few welds that had three or more Arc Starts in a single weld bead and some of them were less than three inches long.

Some robots need to have a separate Arc End for each Arc Start and others need only one at the end of the weld but I have never saw one where it hurt to have separate Arc Ends so I always put one in the program for each of the Arc Starts. These Arc Ends do not need to be at specific points in the weld or program as the Arc Start commands did, in fact, all of them can be piled at the end of the weld, one directly after the other. The illustration below shows the proper way to nest the Arc Starts and Arc Ends in a program. It's very similar to the way brackets are handled in a math problem with the innermost sets being completed first.

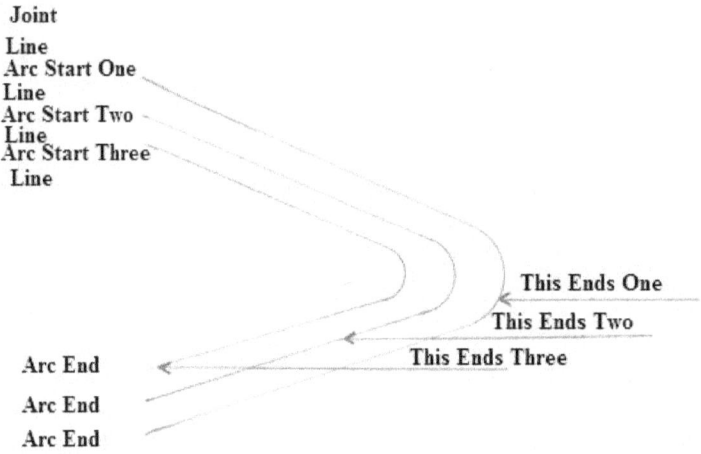

Weaves can also be nested in a program the same way that you nest the Arc Starts. If you need to change to a different weave type in the middle of a weld, all you have to do is just add a line or circle, which ever you need to tell the robot where to change the weave condition. Then add another Weave Start that has the different parameters in place. Just like with the Arc Starts you can use as many different weaves as you need during the course of a weld. I also follow the same rule with weave ends too, one for each weave all piled up at the end of the weld.

With some robots, like the newest OTC Daihen models, the weaves only work during the weld. If you are lucky enough to have one of these robots you are going to have a lot less lines in a program with weaves. With these modern beauties you only have to add a weave start to the program at the first weld that you need weaved or if you want to change to a different weave pattern. These robots also only need a weave end if you want to stop weaving or at the end of the program. Every weld in between the weave start and the weave end will have a weave in it but the moves in between the welds will not be weaving.

On the other hand, if you are on a brand or model of robot that doesn't have this improvement you will need to put a weave start and weave end at the beginning and end of each and every weld, that's a lot of extra work.

It is easy to check to see if your robot has this handy improvement. All you have to do is program a weave into a weld and then back up and check go through the weld and into the points on the other side. If the robot weaves as it goes through the weld but also weaves through the open air movements that are in between the welds as well, then you have an older model and you will have to begin and end each of the welds with separate weave commands.

Circles really should be called a Radius or an Arc because although you can and occasionally do use them for complete circles most of the time they are used for partial circles or to simply wrap corners more effectively. Any curved weld that can be said to have a radius is accomplished through the use of the Circle command.

The rule you have to remember with the circle command is that they must be used in at least pairs; circle 1 and circle 2, or C1 and C2. This is actually just a minimum requirement needed for a perfect circle or radius. The problem with that is that only a very few of the times that

I use the circles are on anything close to a perfect radius. The more variance that a radius or circle has in it, the more circle commands that you will need to use. When you use three or more circle commands the first two will always be called C1 and C2 but if you use any more after that, all of the rest will also be considered to be C2's.

The worst case scenarios are ovals and spirals, with these two shapes the radius never remains the same. Below you'll find three illustrations of ways to use the circle commands.

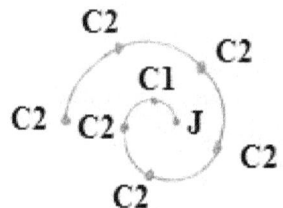

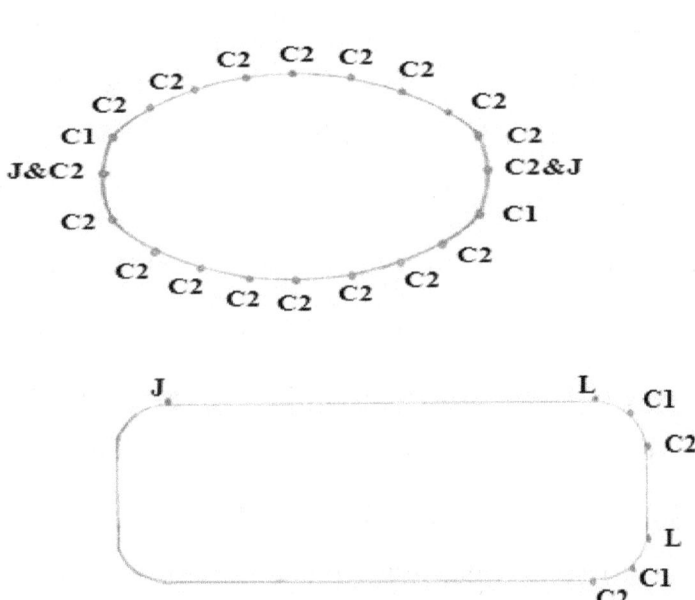

If you write a program and the circle commands do not make the weld follow the radial seam exactly as it should then you need to go to where the problem is and create another circle command in between the two points that are on either side of the problem area. If necessary you can keep repeating this until the robot is forced to follow the path of the radius on the part. You can put as many circles as needed to correct the path of the robot however, if you are having trouble correcting by the addition of more circle commands there may be another problem.

Another common problem that can cause the robot not to follow the correct path around a radius is the angular orientation of the torch. To prevent this from happening you should attempt, when the fixture and part allows, keeping the same angle of attack as you go around the radius.

Occasionally you may find the need to reverse directions of a radius, such as an 'S' shaped weld. This can be done through the use of a Line command. You are not actually reversing directions of a radius even though that is the end result. By putting a Line command into the program you are effectively ending one radius and beginning another. The line needs to be real close to the last C2 and it will serve as the beginning point of the next radius. In the illustration below you'll see how to reverse the radius using the Line command.

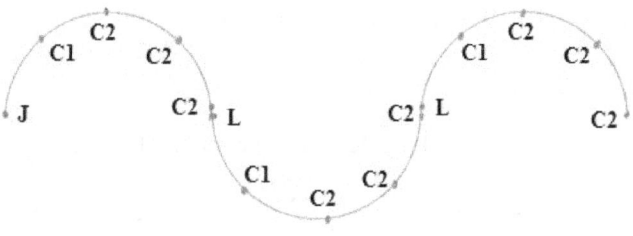

Circle commands are also used vertically as well as horizontally such as when welding a rod or round tube into a piece of angle iron or onto an upright plate or bar. The illustrations below show how this is accomplished.

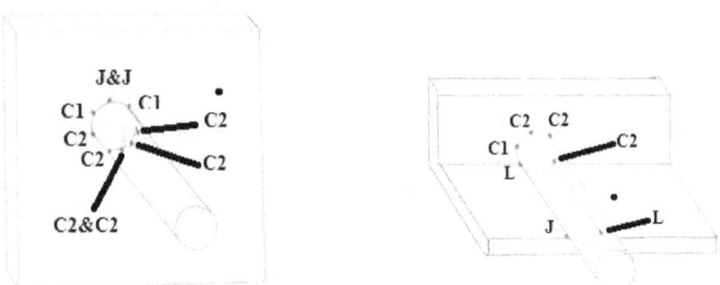

When programming multiple, connecting welds it is possible to run them as a single continuous weld bead instead of separate welds. This is not only good cosmetically but it will also increase the strength of the weld by a significant margin.

This accomplished through the use of additional Line commands inserted in between the Arc Start and the Arc End. Simply place a Line command everywhere you need to switch directions, by doing this you are effectively creating a corner. Each additional Line command serves not only as the end point of the previous weld bead but as the starting point of the next one so long as you need only slight changes in the angle of attack.

The illustration below shows how this is done.

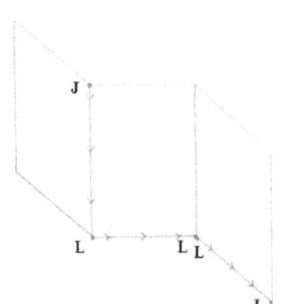

If you need more drastic changes in the angle of attack when doing multiple beads then you'll need to insert two Line commands at or close to the same point on the corner in which the first will end the previous bead with one angle of attack and the second will start the next bead with a different angle of attack but with very little travelling between the two points.

The end result will be that the robot will pause for a fraction of a second while the arc continues as the torch changes to the next angle of attack and then begins the next weld bead, still creating a continuous weld around the corner.

The illustration below shows how this is done.

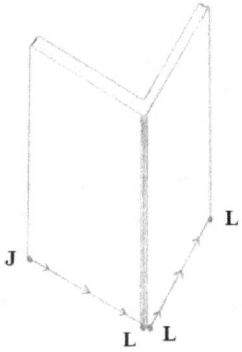

Plug welds are different than most other welds because not only do you have to burn in and penetrate to the plate underneath you have to fill the hole and make sure that the edges are not undercut. This is done because most of the time the customer wants the plugs to be sanded down effectively disguising their existence. When properly welded and sanded flush a plug weld will end up being the same as a spot weld. Plug welds are done for a great variety of reasons but the two most common of those reasons are when the company doesn't own a spot welder or the metal is too thick to be spot welded.

How you weld them depends of several variables; the size of the hole to be plugged, the thickness of both the outer and inner materials being plug welded and the position of the plug weld; flat, vertical or on an angle.

When plugging the smaller holes many times you can get by using only a tack weld. Come in straight to the center of the hole, if possible you need to be completely perpendicular to the surface of the plate. If you can't, then you need to get is as close as possible. If the plug weld is on a vertical surface it is sometimes better to have the torch angled upwards a couple of degrees or so.

The settings for a plug weld are usually a little higher than if you are welding a seam or joint, this is because that you want to make sure that it gets good penetration into the base plate before the hole begins to fill in.

If the robot is doing a small plug weld in the flat position you can usually allow the hole to fill completely using only one long tack weld but with any other position you may have to use multiple tacks. This is done without moving the torch or changing the angle of attack, using only the timer function.

When you program the robot to make a small plug weld, you need to first line up with the hole and have the end of the nozzle about two inches away from the work. Next you need to move the torch head straight in making sure that the tip of the wire is dead center of the hole. Set the amps in the Arc Start condition file about fifty percent higher than you would for a regular weld with those same materials, the travel speed doesn't matter because being a tack weld it isn't going to move.

Always create a new condition file so if you do need to change the amps you can just change the file and not hunt another one with the right amps. Make your best guess for the length of the tack weld and set the timer, you'll probably have to go back in and change that as well.

This is yet another case that demonstrates the need for a robot programmer to also be a good welder, that experience will always give a programmer a head start when it comes to making an 'Educated' guess at both the weld conditions and timer settings.

After setting the heat and length of the tack weld all that's left to do is put an Arc End after the timer to end the tack.

After programming it you'll need to test the weld, most plug welds will need adjustments especially when you first begin doing them. Even now, unless it's identical to others that I've done in the past, I'll still end up changing the amps or timer if not both.

There are several things to look for when you test a plug weld. One of the most important things to examine is the heat. Both the base plate and the outer plate surrounding the hole that's being plugged needs to turn cherry red during the weld to show that the tack weld is penetrating enough. If you don't or can't watch it as it is being welded a close examination of the plug weld afterwards should show sufficient discoloration of the metal on front and back of the plug to indicate that the penetration occurred.

Next, check to see if it filled up the hole completely, if there is any doubt, just sand it flush, there should be no indication that there was ever a hole there to start with. If the plug weld did not fill the hole up then you'll need to add more weld. On a flat weld you'll probably just need to increase the length of the weld by increasing the length of the timer. If the plug weld is in any other position you will likely need to add a timer after the Arc End to add a pause allowing the weld to cool slightly before using another tack weld to add volume to the weld. There have even been instances where I have actually had to use a third tack to insure that the hole is completely filled in.

If you are plug welding a hole that is in a vertical position you'll need to check and see if the tack is drooped down because of gravity. If this is the case you may need to shorten the length of the tacks and add more of them to prevent this from happening. If you are using multiple tacks you may also need to lower the amps on the subsequent buildup tacks, it's usually a good idea to do this anyway because after the metal is already hot you shouldn't need as many amps and the additional tacks are only for adding material.

If you can't get sufficient penetration into the base plate even after raising the amps on the tack weld then the problem may not lie in the weld but because the hole was too small. Occasionally I've had to send a project back to the drawing board because the hole was simply not big enough to allow the weld to penetrate before the hole in the outer layer filled up.

On larger plug welds you'll have to use more than just a tack weld to achieve the desired results, these often will require you to program a four or five point circle. You'll need to establish a point that lines up with the center of the hole with the end of the nozzle about two inches away from the work as you did with the small plug welds. Next, come in and start the arc in the direct center of the hole. Unlike the smaller plug welds this time you will use an amp setting close to the same and you would a regular joint or seam weld. Then using the timer function, pause for a half second or so to give the heat time to build up slightly before moving to a point close to the top edge of the hole. That second point will need to be a Line command because you are going straight from the center and it will be the first point of the circle. After that you will need three to four circle commands, I prefer to use four; C1, C2, C2 and a final C2 that is at the exact same spot as the Line command at the top of the plug. Each of the circle commands need to be as close as possible to ninety degrees apart. I have also found it best to angle the torch in towards the edge of the hole all the way around the circle, three to five degrees is usually enough.

The illustration below will give you a visual description of the way to program a larger plug weld.

After you test weld these larger plug welds you will need to look for exactly the same things as you did when inspecting the smaller plug welds. The main difference between the larger plug welds and the smaller ones is in how you tune them in. With the smaller plug welds you mostly adjust the timer but these welds are in motion so you will need to adjust the travel speed instead.

If the plug weld is in the flat position and the metal is thick enough to withstand the extra heat then all that you usually have to do is slow the travel speed down enough to allow some more weld to build up. If it is too thin for that you may have to use a timer to pause and allow the plug to cool and then use multiple passes to add sufficient material to fill the plug up.

On the other hand, if you are doing a vertical plug weld slowing the travel speed down may cause the weld to droop so you may need to do multiple passes instead. Again you may also need to put a pause in between the passes to allow the previous pass to cool slightly.

When filling in these slightly larger holes you sometimes have to end up with a tack in the center to complete the backfill. The heat and duration of the passes and/or tacks needed to fill the plug will be determined by the size of the hole that is to be plugged, the thickness of the material being plugged and the orientation of the weld. Once you have the penetration into the base material and the connection to the top plate then all you need to do is fill up the hole by any means needed or possible.

For even larger holes and/or thicker materials you may find that you will need to use graduating circles each one progressively smaller than the one before it to backfill the hole, as shown in the illustration below.

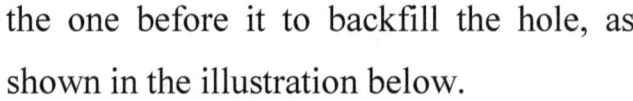

The amount of inner circles needed as well as the amps, travel speed and cooling times will always need to be determined through a combination of trial, failure and experience. If the plug welds are this big then you may even need a second run of these graduating circles to achieve the amount of backfilling required. Holes that need this much weld are actually far too large and take far more production time to plug and sand down which in turn eats into the profit line. I suggest going back to the engineering department and asking them to reduce the size of the hole so the plug welds can be completed in a more reasonable amount of time along with being much more cost effective. If it's the customer or the engineer that wants more weld attaching the two plates together then the best way to achieve this is either with more of the smaller holes or by using slots instead. I wish you luck with the engineers, more than I usually have any way. If you can find a way to make them think that it is their idea you'll stand a better chance. Getting an engineer to admit that they made a mistake is a rather difficult endeavor.

A proper sized slot is usually programmed with two or three passes. Begin on one end followed immediately with a short pause to let the heat buildup before heading for the other end of the slot and then end with another short pause. On the first pass, which should always be on the bottom edge of the slot, I always have the torched angled down a few degrees. The second pass should be on the top edge of the slot and on that one I always angle the torch up slightly. The second pass doesn't usually require the pauses. If you do need a third pass in the center then keep the torch angled up for that one as well. In the illustration below you'll notice that the slots are usually programmed using Line commands instead of circles. As with the large plug welds, if the slot takes more than three passes to fill then it is also too large.

The biggest part of learning to program a welding robot is the gaining of experience in moving the Robot inside of the X, Y, Z Space that is the robot cell while maneuvering around the various fixtures and parts. You also have to do this without surpassing the limitations built into the robot. This is mostly accomplished by trial and failure and through repetition with the robot screaming at you telling you that it can't do that.

You will find however, that the old adage or maxim 'Practice makes Perfect' doesn't apply where Robots are concerned. While it may be true that the more experience that you manage to acquire manipulating a robot arm, the better you will get at it, there is no such thing as perfection, there is always room for improvement.

Many of the orientations that you will attempt to get the robot into, or the direction that you will try to force the torch to weld in, will not be achievable and the robot will unwaveringly refuse. It will tell you, through error messages, that it is impossible and will steadfastly refuse to even try. Although this will be extremely frustrating it is your job, as a robot programmer, to find ways around that apparent stubbornness. It is sometimes not easy but you'll find that most movements and positions are actually possible; you simply have to be more stubborn than the robot. There will be times when you will find yourself using nearly every motion key on the teach pendant just to get between two or three points especially when you are getting into or out of tight places in between the fixture and/or parts. Many times I have been reduced to setting the robot in the mode where I am only moving one joint at a time, trying to fine tune the orientation so that I can get something to work.

You may even find that you'll have to go back and add even more points just to keep the robot from crashing into something. Your rotational keys are usually your best asset in your arsenal of tricks. When using the standard robot coordinates you will eventually be able to predict which way the arm will flex but when it is set in the tool coordinates there are far too many factors involved for an educated guess. The position that the arm is in will make a huge difference in the way it moves in this setting. So when you're in one of those tight spots just start trying the rotational keys in both coordinate settings until you find one that gives you the desired orientation without changing the angle of attack, or at least no more than necessary.

Manipulating the robot arm to get it where you need it is one of the times when you will simply have to show the robot who is the boss. Resistance is the Key and Stubbornness is the method to success; NEVER surrender to the robot.

Some of my coworkers have occasionally gotten a good laugh if they happen to walk past when I'm arguing with the robot. This is because I will sometimes be fussing at it just as if it were a real person. I've even had nicknames for all of the robots that I have ever programmed; this is because all of them seem to have different personalities that unfailingly come out just when they are being stubborn about something of this nature. Another reason for me to give them the nicknames is so that I can address them properly when I'm cursing at them with a handful of somewhat colorful metaphors.

When programming a fixture that has several identical welds in a row that have to be welded all the way around like either square, rectangular or round tubes, I have found that it is usually NOT in your best interest to complete each of the welds individually as shown in the illustrations below that are labeled 'The Wrong Way'.

This is because it will take much less time and effort if you will instead program all of the welds that need the same orientation of the torch head one after the other as shown in the illustrations below that are labeled 'The Right Way'.

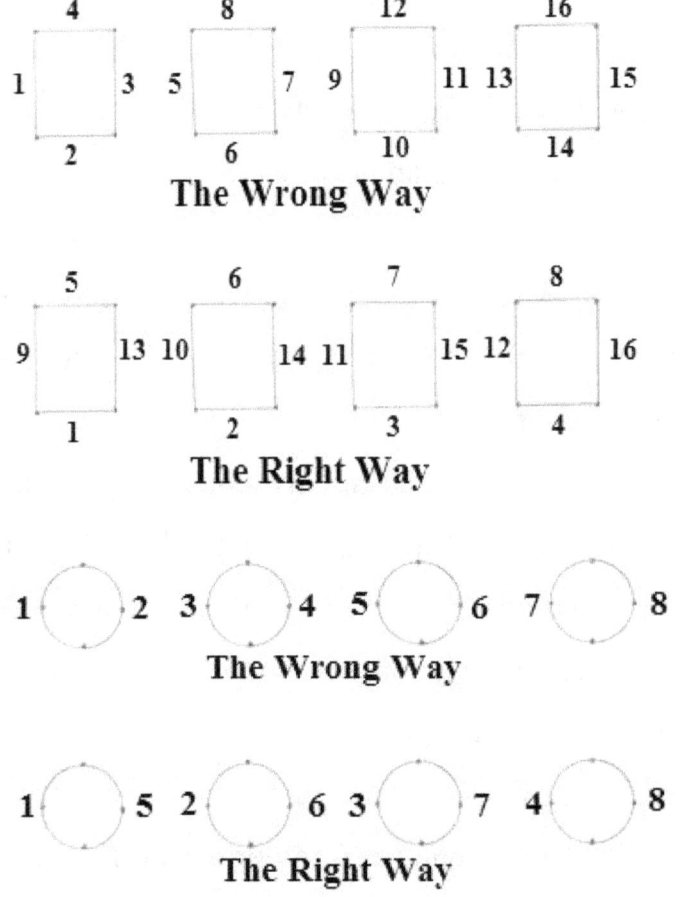

Once you have the torch head in the correct angular position for a weld, which is one of the hardest parts of your job as programmer, you should complete all of the welds that need that same orientation of the torch head or angle of attack.

Not only will this save you a lot of time and effort, which in itself is a desirable effect, it will also produce a cleaner program that has a lot less steps and reorientation moves in it. This translates into a program that will run much faster and more efficient as well as producing welds that are more consistent and a product with a much better appearance. This will also have the desirable effect of making your boss, and their boss, right up to the top of the ladder, very happy. Then along with all that there is the added bonus of sliding another feather into your hat which is never a bad thing. One thing that you always need to remember; if the boss ain't happy then nobody underneath is happy either, and most of the time this will usually include you.

Troubleshooting the Programs

A robot is not supposed to be able to forget where any of its programmed points are located, ever, that unerring consistency is one of their largest selling points. That being said please DO expect it to happen from time to time, because it certainly will. Every robot that I have had the pleasure of programming has had a touch of amnesia on occasion.

Don't laugh, I know it sounds funny but it's not a joke. Sometimes, especially when troubleshooting, it helps to think of these robots as a coworker, they will sure act like one.

Most of the times that this apparent amnesia occurs will be when you put an old fixture back on the robot, especially if it is one that hasn't been run in a while. The chance of this happening is what made me decide to never turn a robot loose running a program without at least a preliminary check to make sure that it can get to the beginning point of the first weld.

There will also be times that a program will be running perfectly when this amnesia will occur. When it happens like this, it will NOT come on gradually, it will be sudden and without warning, usually between one run and the next. All you can do is shake your head and go in and fix it. This is what some people refer to as an example of the Extension to Murphy's Law; "Even if it CAN'T, it might."

Once, when I was working with a rather old robot, I'd jokingly refer to this as Alzheimer's, and there were times that I would wonder if it was actually a joke. I mentioned earlier how I had given every robot that I have worked with nicknames according to their individual personalities, this old fella I named 'Jetfire' after the old broken down Transformer from the movies of the same name. In the next room was a brand new robot and naturally he became 'Optimus Prime'.

If this seems to happen but instead of happening suddenly it comes on gradually then it is almost always caused by something else; contact tip, nozzle, the wire, component parts, etc…NEVER change the program without checking everything else first.

The very first thing that I always do, at the first sign of trouble, any kind of trouble, regardless of the symptoms, is to change the contact tip. You will probably be surprised as you discover just how many different problems with a weld will be corrected by this simple little action. I'm serious, sometimes as much as seventy-five percent of the time or higher you'll find that the contact tip is the whole cause of the problem.

Contact tips are made of copper, a very soft metal even when cold, and the heat generated by the welding process can easily warp and disfigure them. A small piece of weld spatter stuck to the end can force the wire to come out crooked which will cause the weld bead to end up in the wrong place. That can be very deceptive giving the appearance that the programming is off, don't be fooled. A contact tip can also appear to be perfectly fine but still be the cause of the problem. Do not trust appearances; change the tip just to be sure before doing anything else, they don't cost that much and you can always put the old one back in if the change made no difference.

Another reason that I always suspect the contact tip is because of a regularly occurring problem that is completely invisible. As the tips heat up they get even softer than usual while the steel wire that is going through the tip remains cold and hard, it is also curved because of being housed either in a drum or on a spool. As that curved wire is pushed through the softened copper tip it is constantly rubbing on one side of the hole. The wearing action allows the wire to retain more and more of its curved shape. This is another problem with the contact tip that can make it appear that the programming has changed moving the weld from its recorded location. The really bad part of this problem is that it is completely invisible because it is happening totally inside the contact tip itself, there is no change in the outward appearance of the tip when this happens. The only indication that anything is amiss is when the weld appears to move away from its programmed location and it happens so slowly that by the time it's usually noticed there have already been quite a few welds that have been adversely affected. You should always suspect a problem with the contact tip first; it will save you a great deal of time and trouble in the long run.

Another consumable that will often cause the weld to wind up in the wrong location is the diffuser, sometimes referred to as a 'Tip Body' by many manufacturers. They are usually constructed of either brass or copper, both of which are soft metals that are superb conductors of heat. Along with being excellent conductors of heat they are also easily deformed by the intense heat of the welding arc. You always need to keep a check on this as you clean the nozzle. If a diffuser is warping it will invariably result in the end of the contact tip not being in the center of the nozzle, very easy to spot if you are looking for it so you need to always be on the lookout for it.

If you see that the contact tip is no longer centered and the nozzle itself is good and tight, then there are only two things that can cause this; either the contact tip or the diffuser is warped. Always put in a new contact tip first and if that doesn't correct the problem then change the diffuser. Always take care of this immediately after you spot it even if it hasn't made a great difference in the welds yet. It is already affecting them whether you can visibly see the difference or not and it is going to always get worse, do not procrastinate, it will only cost you more in the end.

It's a good regimen to always check for this problem before tweaking the points of a program. Before getting into this habit I tweaked several programs to put the points back into their proper locations only to have to move the points back after changing the diffuser. After a while you begin to get tired of kicking yourself in the butt, always check first. It takes a lot longer to correct a program that you altered unnecessarily than it does to make sure that the tip is still centered, this is yet another case of being better safe than sorry.

Another thing that can cause the weld to be out of place is running the wrong size of contact tip. The proper size contact tip is essential to producing good weld bead anyway so never run one that is the wrong size. They are called contact tips for a good reason; they need to be in contact with the wire all the way through the tip. A contact tip that is too large will allow the wire to retain too much of the curved shape as it travels through it which in turn allows the weld to move.

Yet another problem that can also cause the weld to end up in the wrong location is one of the liners; either the shorter one between the wire feeder and the torch or the longer one between the drum and the feeder. This is not something that happens often but should never be overlooked when trying to figure out what is wrong with a weld. Either of the liners can pick up trash, if this happens in the short one immediately before the torch it can cause the wire to reform and curve in a different direction. If it does this it results in the same outcome as if the contact tip or diffuser is warped; moving the weld. Trash buildup in either liner can also cause drag on the wire which can in turn cause a multitude of weld problems such as thin, underdeveloped beads, uneven or skipped places in the welds and can even cause the wire to melt onto the end of the contact tip.

I always check the liners for drag at least once a month or whenever a problem arises. To do this is simply release the spool(s) of the wire feeder and pull a foot or so of the wire through by hand. The wire should pull through smoothly and with very little effort. If you feel any resistance at all you will then need to cut the wire just behind the wire feeder and try each one separately to determine which liner is giving you the problem and needs replacing.

To reduce the instances of this happening and therefore lower the overall operating cost of the robot you should always keep a clean and lubricated felt wire cleaner just inside the drum. These can be purchased from any welding supply company; you can get either the pre-lubricated cleaners or you can get the dry cleaners and a can of the lube and lubricate it yourself. This is a low cost way to avoid buying quite so many of the more expensive liners, a common sense preventative maintenance solution that should be used on all MIG welders manual or robotic.

Another thing that can adversely affect a weld is a ceramic baffle, sometimes referred to as a diffuser cup. Most robots that have them will operate just fine without them. I have never personally used them since finding out that they are not needed and will, to a large degree, adversely affect the weld if damaged. In fact, I have never found a single case where using them added to the quality of the welds produced. However, if you are determined to continue using them or are being forced to use them by the owner of your robot, change them at the first hint of a gas related problem. A tiny hairline fracture so small that you absolutely cannot see it can cause the weld bead to be so full of porosity that I have actually seen them broken by hand.

These little ceramic cups, that are so easily broken and serve no useful purpose, are actually a very ingenious invention. Their sole apparent purpose is to give the manufacturer another consumable to sell you, one that you are going to have to replace quite often and actually cost a lot more than some of the others do.

One of the most common reasons for a misplaced weld hasn't got anything to do with the robot itself and is also one of the most commonly overlooked problems as well. If a robot is welding right beside the joint always check to see if the material is cut and/or machined correctly. If the joint is in the wrong place, however slightly, and the robot is welding in the right place then it will appear that the program is off. Another simple problem that will commonly give the same result is weld spatter or a buildup of grease and trash on the fixture; always keep the fixture cleaned off in between the runs. A good air gun and a rag that are used regularly can prevent this problem from ever occurring.

If everything else checks out okay then you will have to go into the program and tweak the points to put the weld back into its proper location. I always leave the tweaking until after I am sure that it is not a consumable that needs changing, component parts that are wrong or a fixture that needs cleaning. There is nothing worse than modifying a program and then later finding out that you have to go back in and change it all back after you find the real cause of the glitch, especially if it was just a simple and sometimes obvious remedy to the problem.

If you write a program using circles but when you check through or test run the program you notice that the torch seems to speed up or slow down as it goes through the circle commands you will need to tweak the program so that it will run smoothly. If you don't correct the problem the weld will get thinner as it speeds up and fatter as it slows down which will ruin the appearance of the weld as well as reducing the strength significantly. If you get the robot to weld around the radius smoothly the weld bead should be consistent from beginning to end. The illustrations below show the right and wrong way to program a circle; the angle of attack is the critical factor:

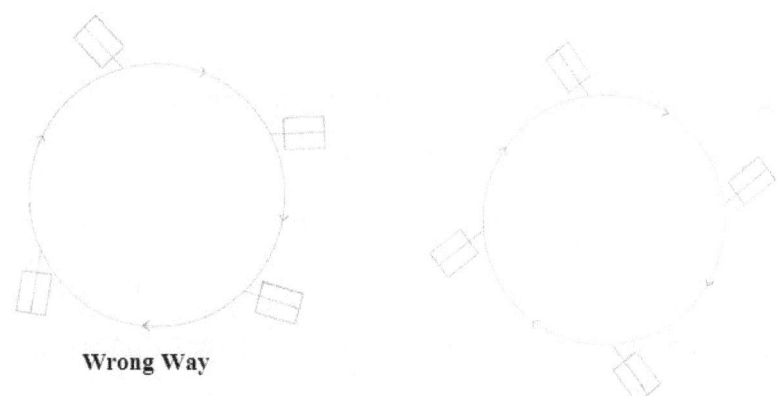

Wrong Way

Right Way

Inconsistent motion around a radius is usually caused by swinging the torch head too much or too little between two of the circle commands. The angle of attack should be held exactly the same all the way through the radius if possible. If it is not possible to hold a consistent angle of attack, then just aim to keep it as close as possible. The best way to correct this problem is to adjust the angle of attack of the points before and after the problem area. First jog the robot to where the inconsistent motion is taking place; if the motion between the points is slowing down you'll need to add some more motion to the first point and remove some of it from the second point. You may even have to add another point before the first point or after the second each with a small amount of the motion to split it into smaller segments. If the motion is speeding up between the points you will probably need to remove some of the motion from the first point and add some to the second. If correcting this is unfeasible or impossible because of a danger of collision with either the fixture or the parts themselves then you may need to add a nested Arc Start/Arc End to change the travel speed of the weld to cure the symptoms of the problem instead. In other words, if you can't raise the bridge then sometimes you just have to lower the river.

When programming a robot you sometimes have to get a little creative, especially when you are attempting to find a way around a problem. I have been told on occasion that the methods that I sometimes employ to achieve the results that I attain are a little unorthodox, but when dealing with a robot I have found that sometimes the ends do actually justify the means, if it works then it's a practical solution to the problem.

There are several things that can adversely affect the welds that although they don't have anything to do with the weld conditions it's those very settings that I sometimes need to adjust to correct the problem.

One that I've had to deal with occasionally is when the company would change to a different wire. I'm not talking about a different type or style of wire; I am speaking about the same wire from a different manufacturer or supplier. All welding wire of a specific type is supposed to be the same regardless of who is making it, don't believe it, it's a lie. There have been several occasions that I have had to change the amps and/or travel speeds to get the same weld with a different name brand wire. One company that I worked for found a wire that was not only supposed to be the same wire as what we had been using, it was being advertised as a higher quality welding wire. We tried everything that we could think of to get the wire to produce welds as good as what we were getting with a 'lower quality' wire. The engineers at the company that produced the wire told us some different adjustments to make and we tried all of them, to no avail. The finally tried to tell us that there was something wrong with our robot, that's when the owner of our company told them to come get the trash wire that they were trying to sell us and we went back to the wire that we had been using, which the robot still ran just fine. After inquiring a little deeper I discovered that the new and improved wire that we were trying to get the robot into running was going to cost less than half of what the company was currently using, little wonder that they really wanted it to work. This turned out to be yet another case of someone getting exactly what they were paying for.

One would think that if a company paid, as that one did, a six figure price for a brand new, top-of-the-line robotic welding system that they would want to have a top-of-the-line welding wire being used in it.

I have run head on into two more instances of a low-cost wire causing a problem that the robot simply couldn't handle. In one of those cases it turned out that the wire was two thousandths of an inch thinner than it was supposed to be. I know that does not sound like a significant amount but it was enough for the wire to lose contact with the 'Contact' tip and that's what was causing the problem. The other instance was with a wire that no one could find a problem with; it caused the welds to have a significant amount of porosity. Finally I noticed that the wire just seemed to look just a little too bright, almost like a new copper penny. That's what gave me the idea to take a portion of the wire and use a piece of Scotch-Brite pad to clean it with. All this did was remove a small amount of the copper coating from the wire but when the robot used that section of wire the welds suddenly began to run right again.

Another thing that has given me problems of this same nature is with the shielding gas. Slight differences with the mixtures can make a huge difference in the weld settings that you will need to produce the same welds. If your company is still using bottled gas then I strongly suggest that you switch to a system where you mix your own gas at your place of business. This gives you much more control of the mixture that is getting to the robot. With bottled gas you get what the supplier gives you and most of them have a two percent cushion in each direction, this means that if they are anywhere within that four percent range that they can't be held accountable. A four percent difference in the mixture to a robot welder is enough to ruin the welds produced.

If you live in an area where there are severe temperature differences between the seasons that can also make a difference in the gas as well. For one thing the gas tends to concentrate during the cold weather months so that you are actually getting more gas per cubic foot than you are in warmer weather at the same pressure. Some of the gasses tend to concentrate more than others which will have an end result in changing the mixture percentages as well.

Another gas related problem relative to the weather is settling. Gasses have different relative weights which can cause the gasses to separate in what can be referred to as the centrifugal effect. Holidays are the main cause for this problem when the robot or the entire plant is shut down for three or four days in a row. This is more common with bottled gas but can also take place to a smaller degree in a piped in system. If you suspect this has happened the best way to cure the problem or eliminate it from the list of possible causes is to do a gas purge.

Most robots have a gas key that is made for this purpose but I usually don't use it. I have found that it is much simpler and more effective to just loosen the gas line from the regulator a bit and let the line purge itself. This is much better to me than holding the key in for a few minutes straight.

There have also been occasions where the composition of the metal itself changed either through simple mill-specs, the customer ordering the change or the company finding a cheaper material from another supplier, which in my experience the latter is actually the most common cause of this occurrence. For whatever reason the change was made, different materials sometime melt at slightly different temperatures and will cause you to have to go into the programming and change the welding conditions.

If nothing else works then you may have to change the weld conditions. This is the absolute last resort because if a weld was running fine before then you really shouldn't have to make adjustments to the weld conditions to get it to run well again. That being said, it does occasionally happen.

You will also, as you gain experience, run into times that you will never know what happened but if you have tried everything else then adjusting the weld conditions is the only other alternative. In a case like this I will make a new Arc Start file with the new conditions instead of changing the condition that is in place. Then I will create a comment line describing the change that I made and why I did it. This way, if I need to change it back later when and if the situation reverses then I'll know exactly what was working before. The beauty of comment lines is that they have absolutely no impact on the program because the robot ignores them; they are only for you, as the programmer, to read.

Note: Make comment lines for yourself anytime that you think that you may need to be reminded of something later, it can be a big help.

OTC Daihen

This chapter is for those of you that happen to be lucky enough to be learning to program an OTC Daihen robot. I'm not saying that the information in this chapter will not apply to other robots because most of it probably will, it's just that the commands that I'll be using will likely be different on those other robots. These are my robots of choice and the ones that I have the most experience with as well as being the robots that I use to train potential programmers on so I am dedicating this chapter to the user specific instructions and commands that the programmers will need to learn if they are going to be using the OTC Daihen brand robot welders.

The teach pendant of an OTC Daihen is easy to use and safer than the ones belonging to many other robots, especially on the newer models. The new and improved dead man switches are 'Three Position' switches. What this means is that you have to keep them in the exact position for the robot to move, this is what I commonly refer to as the

'Sweet Zone'; not too tight, not too loose, it has to be just the right amount of pressure.

Their R & D department finally figured out that if you are getting pinned by a robot that a human's reaction is sometimes to tighten up instead of letting go. With the older models as well as many of the other brands the dead man switch remains on as long as you are squeezing it and you may keep right on squeezing it until the moment that you die. I realize that this is the original meaning of the term 'Dead Man's Switch' that it releases at the moment of death, but that should not be the objective here.

With the improved OTC Daihen dead man switches you have to keep the switch in the 'Sweet Spot' in the middle. Too tight, not tight enough, both depressed at the same time will all bring the robot arm to a screeching halt. Your fingers may need to take a break once in a while but if it saves your life it is well worth a little time lost or discomfort. With some other name brand or the older OTC Daihen pendants I've saw programmers that actually wedged a penny or a dime into the crack beside the switch to lock it in the on position so they wouldn't have to hold it at all, what good is a dead man's switch if it won't even turn loose after you are dead.

If you are also lucky enough to have one of the newer FD teaching pendants then you are not going to get quite as tired as you would have with one of the older models. The FD pendant is nearly half the size and about twenty-five percent lighter which makes programming easier and more comfortable. They have simplified the backup of all of your data directly from the teach pendant itself by the addition of a USB memory slot so that you can now save all of your data onto a flash drive.

This USB slot is the latest addition and I haven't had a chance to try it out yet but if you happen to have two or more identical robots with this option you may be able to switch programs from robot to robot. I realize that there would be slight differences between so called 'Identical' robots but even if you needed to do a little tweaking it would still be a great time and work saver especially with any of the particularly large programs.

They have also added a welding condition database that will aid the programmers by suggesting welding conditions based on the plate thickness, joint type and the travel speed. Once these presets are input the AMPs will be automatically calculated. This function can help you achieve the optimal welding conditions much easier and faster than was possible with the previous models. You still have to fine tune the welds but will at least start out in the right ballpark. This is especially true when you switch to a material thickness that you are not used to setting up a weld for or a material type you are not familiar with.

The FD pendant also has a new Jog dial roller wheel that makes it easier to scroll through the programs and even makes inching and retracting the wire much easier as well. And there is a new One-Touch Access that will reduce the amount of keystrokes required to navigate through the menus.

Before you even get started you will need to know what most of the keys on the teach pendant are used for. In this next section I'm going to go over these keys and give you a short introduction to what each one of them does and/or when you need to use them.

The first thing that I want you to learn about is the green 'Enable' key. A few other keys on the teach pendant are either fully green or half green, and can be used in conjunction with the Enable key. If a key is all green then, with a couple exceptions, they have no function when pressed on their own. Many of the other keys are half green; with these the Enable key works like the shift key on a keyboard, and you must press and hold the Enable key in to make use of the green part of the key.

The Enable key also changes the nature of certain function keys (F1 – F12); some of the icons only appear when it is pressed, others disappear and other icons that are grayed out become usable when used along with the Enable key.

The second thing that I want you to learn is how to use the help section that comes with all of the new windows based teach pendants. The entire robot operating handbook is stored inside the help section of the operating system. I knew a programmer once that actually learned how to program a robot with little more than the information contained in there. In fact, many of the tricks that I have learned over the years came from playing around in the help section. When I was first starting out as a programmer, I spent quite a few of my lunch breaks exploring around in there myself. There's a lot of good information hidden in there, stuff that you don't see taught anywhere else.

Simply pushing the key that is labeled 'HELP' will open up the section for you. The four arrow keys will enable you to scroll through the contents menu. Holding the Enable key while you depress the right arrow key will switch you over to the index menu and the search function, Enable and the left arrow key will return you to the contents.

When you are in the index you can press the edit key to bring up the soft keyboard and type in what you are looking for. The key with two windows and an arrow between them will switch you to the main window so you can scroll through it and read that section of the book, when you press it again will return you to the content/index side. After you finish exploring the help section just press the Reset/R key to exit out to the regular screen again.

The 'Interp/Coord' key when pressed on its own will switch to a different coordinates system; Joint, Robot or Tool. This key is one that you will use a lot, it changes the way the robot will move when you use the motion keys. When pressed along with the Enable key it will select the interpolation type; Joint, Line or Circle in the recording status line in the Normal teaching mode.

The 'Check Speed/Teach Speed' key when pressed on its own the manual teaching speed is changed. You will end up using this one even more than the coordinate key. Each time you press this key it will change the speed by one. The speeds are one through five with five being the faster. The change is made in a loop all in one direction so if you want to slow the robot down you have to keep pressing the key until it passes five and goes to one and then comes back to the speed that you desire. When pressed along with the Enable key it will change the operational speed when using the Check Go or Check Back functions. This will not affect the playback speed, which is determined solely by the instructions in the programming. The only way to affect the playback speed is through the override function using the enable key to change the properties of F11 and F12 to override up and down.

Note: If you are trying to move the robot with the Check Go or Check Back function but the robot doesn't appear to be moving check to see if you have accidentally switched this speed down to one. If it hasn't happened already then it will eventually, knowing to check this can keep you from feeling like a fool later when you ask for help, as I did once.

The 'Stop/Continuous' key when pressed on its own it switches the robot from continuous to non-continuous operation during the Check Go or Check Back functions.

Note: During the Check Go function you can press and hold the Enable key and the robot will move continuously through the program until you release the Enable key which will cause the robot to stop at the next point. This will not work in the Check Back operation only when you are moving forward through the program.

When the 'Stop/Continuous' key is pressed along with the Enable key it will stop playback of a program. It has the same function as the Stop button which is much easier to use and more readily available when a program is being played.

The 'Close/Select Screen' key when pushed on its own you will switch between multiple screens that are opened at the same time. This is most used with the Help section or when monitoring robot operations. When used along with the Enable key it will close the current screen or window.

There are twelve Axis Operating or Movement keys; X-, X+, RX-, RX+, Y-, Y+, RY-, RY+, Z-, Z+, RZ- and RZ+. When you press these on their own they have no function at all but used along with the 'Dead Man' switch will cause the robot to move.

The 'Check Back' or 'Check Go' keys when pressed on their own have no function. Both of these must also be used along with the 'Dead Man' switch. These two keys are used to jog the robot through the program either continuously or step by step.

The 'O.Write/Rec' key when pressed on its own during the Normal teaching mode is used to record the movement commands. When it is pressed along with the Enable key a previously recorded movement command will be overwritten.

The 'Ins' when pressed on its own has no function. When pressed along with the Enable key is used to insert a movement command during the Normal teaching mode. The movement command is inserted after the current step.

Note: If you had rather have it set to insert the movement command before the current step you can do so in the Constant settings under; Operation constant, Operation condition to Step insertion position.

The 'Clamp Arc' key when pressed on its own switches the designation of the function keys to the Easy teach mode but has no function when used along with the Enable key.

The 'Mod Position' key has no function when pressed on its own but when used along with the Enable key will modify the current position stored in the movement command to the current robot position.

The 'Help' key will open up the built-in tutorial function for help concerning an operation or function.

The 'Del' key when pressed on its own has no function but when used along with the Enable key the current step is deleted.

Note: On some of the older models it will delete the next line in the programming. When I switched to the newer models this was the hardest change to get used to.

The 'Reset/R' key clears the input or returns the setting screen to its original status. It also enables the R codes or Short Cut codes to be input or selected. If you have gotten into something that you really don't want to be into then this is the key that will usually get you out of it. If you are several screens deep into something like Service Utilities then you may have to use several strokes of this key in a row to get back out again.

The 'Prog/Step' key when pressed on its own is used to go to a specific step inside a program; this is especially useful when maneuvering around in the larger programs, or when switching back and forth to specific lines of multiple programs or from the beginning to the end of a program while putting the curtain dodging moves in place at the end of the program. When used along with the Enable key will bring up the program screen where you may choose to go to a specific program, pull up the directory or list of programs to choose from, copy, delete or rename a program.

The 'Enter' key enters the menu selections or numerical input contents.

Note: If you don't use this key when inputting data the computer will change it back to the default data when you move to the next box.

The 'Arrow' keys pressed on their own moves the cursor. When the arrow keys are pressed along with the Enable key a page is either moved or changed. When you are using the soft keyboard the arrow keys on their own is how you choose the character but when used along with the Enable key it will cause the curser to jump over characters that you wish to leave alone in the comment line.

- On a screen where the settings are arranged on multiple pages, the page is moved.

- On the Editing screen or a list of programs or condition files it will jump several lines at a time, usually a screen page at a time.

- On either Service or Constant setting screen, radio buttons can be selected.

- On the Teach or Playback mode screen, the number of the current step will be changed.

The 'SPD' key is used to set the speed of movement commands when teaching in the Normal teaching mode.

The 'ACC' key is used to set the accuracy of a movement command when teaching in the Normal teaching mode.

The 'End/Timer' key when pressed on its own is a short cut used to enter and record a delay into the program. When used along with the Enable key it will record an End command into the program.

The Numeric and decimal keys are used on their own they input their numerical data. All of these keys have a different function when used along with the Enable key.

- 9 = C, the short cut to the Circle command.

- 8 = L, the short cut to the Line command.

- 7 = P, the short cut to the Joint command.

- 6 = Sensor, commands regarding sensor will be displayed on the 'F' keys (1 – 12)

- 5 = WS/WE, can be used to begin or end a weaving operation when programming, usually in the Normal teaching mode.

- 4 = AS/AE, can be used to begin or end a welding operation when programming, usually in the Normal teaching mode.

- 3 = Redo, restores an Undo operation when writing or editing a program.

- 2 = Off, is mainly used to remove a check from a check box.

- 1 = On, is mainly used to add a check to a check box.

- 0 = +, the positive sign is selected.

- . = -, The negative sign is selected.

The 'BS' key when used on its own the last character is deleted. When used along with the Enable key it Undoes or clears the operation immediately before when writing or editing a program.

The 'FN' key is used to select a function command.

The 'Edit' key opens the program editing screen. In many situations when used along with the Enable key will open up the soft keyboard.

The 'F' keys (1 – 12) are used to select the icons displayed at both sides of the LCD screen. The Enable key changes the icons that are associated with the different function keys or makes them active.

There are two main ways to teach with the OTC Daihen robots; Normal teaching and Easy teaching. Easy teaching is called that for a good reason, it's easier; it also is faster therefore saving you a lot of time when you are writing the program.

When you use the Normal teaching method you use the numerical keys to input the data one item at a time. You work within the status line, the odd colored line at the top of the screen.

To change the interpolation or the type of movement you have to press and hold the Enable key while pressing the key that is labeled 'INTERP/COORD'. Each time this key is pressed the interpolation is changed in the following sequence; Joint, Line, Circle and then repeats.

Next you can set the speed by pressing the key labeled 'SPD', this will open up the modify speed screen. You don't need to press and hold the Enable key to do this because the speed key is all white. After this screen opens you'll need to input the desired speed and press the Enter key.

You can also change the accuracy level. This is the one single advantage that Normal teaching has over Easy teaching, you can adjust the accuracy level as you are creating the point but it's so easy and quick to simply go back at the end of the programming project and change the accuracy levels in the Edit mode that it's not worth the added work involved when using the Normal teaching method. To change the accuracy level when using the Normal teaching method just press the key labeled 'ACC', once again you don't need to press and hold the Enable key to do this because this key is all white as well. Then you just enter the desired accuracy level from one to eight.

After adjusting all of the required parameters in the status line you need to press the key labeled 'O.WRITE/REC' to record the command line. This key is easy to see because it is the only one that is green and orange. You don't need to use the Enable key unless you are using it for the overwrite function which is in the green part of the key.

To do the same thing when using the Easy teaching method takes a lot less time and effort. To change to the Easy teaching method simply press the key labeled 'CLAMP ARC', this will change the twelve function keys to the Easy teach commands. The function keys that are used while using the Easy teaching method are as follows:

F2 – Arc Start

F3 – Arc End

F4 – Weave Start

F5 – Weave End

F6 - End

F7 – Joint

F8 – Line

F9 – Circle

F10 – Wire Inching

F11 – Wire Retracting

F12 – Gas

In Easy teach mode you first decide which interpolation or type of movement that you need and press the appropriate function key; 7, 8 or 9. Set the speed in the screen that opens up and press the F12 function key beside Complete and the command line is recorded into the program. The default accuracy level of eight is used and if you want to change that to a lower level just wait until you are through writing the program and go into edit and change them all at once. That's all there is to using the Easy teach method.

To me it's a no-brainer, I use the Easy teach method any time that I am programming.

After the interpolation function keys, F 7-9, the next most used of the function keys in the Easy teach mode is F2 and F3; Arc Start and Arc End. The F2 or Arc Start function brings up the Arc Start condition file screen. This is where you set up the weld parameters. The file number 0 is the default and it is highlighted, if you know the number of the saved condition file that you want to use all you have to do is type in the number, press the Enter key and then F12 or Complete to record the Arc Start command into the program.

If you don't know the number of the file that you wish to use you can use F8 or Select and it will bring up the list of all the Arc Start condition files that have been created and saved in the past. Scroll down the list and if you find one that can be used for the weld you are programming for now, once the line is highlighted just press Enter and that will choose that file and take you back to the Arc Start condition file screen. F12 or Complete will record that Arc Start into the program.

If none of the files will work you will need to create a new one. The easiest way to do this is to find out the highest number used in the file list and type in the next higher file number into the highlighted file number block on the opening page and press Enter. You will get an error message telling you that the file does not exist, that is of course why you chose that number so just agree by pressing Enter. Now you'll need to press the F11 key or Input Value so that you can revise the default settings and create your own condition file. The first thing that you should do is to put a comment in the box at the bottom of that page. I suggest putting the Amps and Travel speed, this is the description that will come up when you choose Select function or F8 and if you use this information it will help you decide which one you will want to choose. If you choose to give the condition file a name instead of using the suggestion of the heat and travel speed, then you can press and hold the Enable key along with the edit key to bring up the soft keyboard. When you are finished simply use the function key to choose complete.

F9 and F10 is the page up and page down keys, use these to get to the pages where the welding parameters are set. Any setting that you change has to be followed by the Enter key or when you move to the next one the default setting will return. Always use the Enter key to make sure that your new parameters will be recorded. Once you have all or the revised settings in place you will need to press F12 or Complete to record your new Arc Start command line into the program.

F4 and F5 are the function keys that are used for the weave commands; F4 is the Weave Start and F5 is the Weave End. These are used in much the same way as you did with the Arc Start Condition files; Press Select to choose one that you have created in the past or to find out what the highest number is so that you can choose a new number and create a new one. Always remember to press Enter after changing any of the values or they will not stick. Once you have everything set the way you want it press Complete to record it into the program.

F6 is the function that you only use once per program and I never have figured out why they even bothered to put it there. The End/Timer key on the controller serves the same purpose. Just press and hold the Enable key at the same time as the End/Timer key and it will put an End command line into the program. As long as you get an End line into the program it will run, how you get it there doesn't matter.

F10 and F11 control the wire feeder, inching and retracting the wire as needed. Each, when used on its own, feeds and retracts the wire at its normal or Low speed but when either is used along with the Enable key it switches the motors to its High speed. About the only time you'll need the high speed is when threading the wire through the liner as when you are changing drums or spools.

There are many function commands that you can use but there is only a few that you will need often and only a few beyond those that you will use at all. After all of the years that I have been programming welding robots I have still only used a small fraction from the list.

There are two ways to select the function commands; inputting the number directly or selecting it from the list. Either method begins by pressing the key labeled 'FN' at the bottom of the teach pendant.

As long as you remember the function number all that you have to do is to type it in and press the Enter key to record the function line into the program. After you have been programming a while you will know many of the ones that you use often such as (99) or the Comment function. Another function that I use in a great deal of the programs that I right is (41) or the robot Stop function, I use this in conjunction with the previously mentioned cleaning stop that I put into most of my programs. The only other function command that I use frequently is (80) or the program call.

If you don't know the number of the function that you need you will just have to scroll down the list to find it.

The longer that you spend programming, the more often you'll find yourself using the Edit mode. You can make a lot of different changes in there much easier and/or faster than you can in the teaching mode. Deleting one or two lines of code is fast and easy enough in the teaching mode using the Delete/Enable key combination, but this method deletes only one line at a time. In the Edit mode you can delete whole blocks of code in the same amount of time. You can also change the text in comments, accuracy levels, interpolation, speed, etc…just as quick and easy.

Once you open up the edit mode by pressing the edit key at the bottom of the teach pendant, use the arrow keys to navigate the edit screen; the up and down keys will get you to the line that you need to edit. When you get to the targeted line you will need to use the right and left arrows to get to the data that needs to be edited. At that point you'll see directions on the bottom of the screen that will tell you either the limits of change that you can input or what number you need to type in for the change you desire.

Changing the text in the comment lines is usually done through the use of the Soft Keyboard. To bring up the soft keyboard first highlight the comment text that you wish to change and then press and hold the Enable key while pressing the Edit key. Using the arrow keys highlight the character that you wish to input and then press the Enter key. The exception to this is the numbers, you can use the number keys on the teach pendant to directly type in the numbers instead of chasing them around the soft keyboard. If you are using the soft keyboard to edit text in a comment then the comment is automatically displayed. To jump over characters that you want to leave as they are press and hold the enable key while pressing the arrow keys and it will cause the cursor to move to the position where you need it without erasing the other characters. To erase characters place the curser directly in front of that character and press F5 or BS (backspace). Of course you may also use the backspace key on the teach pendant as well to do the same thing.

To leave a blank space in between words or characters you can press either F4 (blank) or F10 (blank) the two functions for spaces, I don't know why there are two of them.

Once you have the comment the way you want it just press F12 or Complete to record the new or changed comment into the program.

One thing that everyone programming an OTC Daihen brand robot needs to know how to do is to change the protection level from user to specialist. The reason you need this is because some of the neatest and most helpful functions will not work as long as you are in user status.

To change the protection level you press the Reset/R key then type in 314. Next press the Enter key and when prompted type in the hard to remember password – 12345 followed by the Enter key again. The robot should at this point inform you that you are now a specialist; I'll bet you thought that it would be much harder than that to become a specialist in robotic programming, huh?

One of the most common uses for a programmer to need to change the protection level is so that you can use the 'Check Weld' function, which all of the newer OTC Daihen robots have. It is one of the best improvements that the research and development department has added to these robots in the last decade, second only to the new and improved dead man switch system which may one day save your life.

The 'Check Weld' function allows the robot to weld while it is still in the teach mode and with the door open so that you can put on a welding shield and closely observe the weld as it is being made. This can be an invaluable tool when tweaking, troubleshooting or just fine tuning one of the welds.

To use the Check Weld first place the robot in the position of the point that is immediately before the weld that you want to check. Change the protection level to specialist and then make sure that the robot is set for continuous motion. If you are in the Easy-Teach mode you will have to press the clamp/arc key to bring the robot out of Easy-Teach and then find the Change Key function beside the F1 function key. Press (Change Key) and the turn on the Check Weld with the F10 function key. Once it's on you should hear a beeping sound, that is the warning letting you know that the next time that you check forward that it will now be welding so you had better protect your eyes.

After double checking that the robot is set for continuous operation, check weld won't work if it's not, put your shield on and then press and hold the check go key to begin the weld. If the weld is burning correctly then let it continue to the end but if it's not then just release the check go key and it will stop welding.

If you have to make changes to the arc start condition or programming the check weld will automatically shut off and will have to be turned back on before you test the weld again. If that weld checks out and you move ahead toward the next weld be sure to manually turn it off before checking through the weld unless you want to check weld that one to. Not remembering to do that has caused me to catch an unwanted arc or two. Also be sure that if you no longer need to be in the specialist mode to turn it off. This is so nobody else can come along and do something that you wouldn't want done and/or don't want to fix. The easiest way to turn it off is to follow the same instructions for turning it on except put the wrong password in, anything will work, and it will put up a message saying that the protection level has been set to user.

If your company has robot operators as well as programmers then you will want to keep all of the programs in the computer protected. Doing this will insure that a non-programmer will not be able to accidentally mess up one of your programs and all of the hard work that went into writing and tweaking it to perfection.

To protect the programs press and hold the Enable key while pushing the function key for service utilities, scroll down to file manager, scroll down to file protect, right arrow over and scroll to all protect, right arrow over and scroll to all programs and then click on Execute. If you need to tweak a program you'll have to release protection from that one program first. To do this all you have to do is go into the service utilities, then file manager, file protect, over to release protection, over to program, type in the program number and click on Execute then hit the Reset/R key until you are back to the program screen. When you open up a program you'll know if it is protected because it will have a red square in the program number box with a one inside. One thing that I have never understood is why you don't need to be in specialist mode to release the protection from a program. It's conceivable that anyone can figure out how to release the protection from a program by simply playing around. Having that function delegated only to the specialist mode would give an extra layer of protection that I personally think is needed. Maybe the R and D guys will realize this and change it in the next generation of robots.

The Reset/R key, as you saw in the last exercise, will back you up to the previous page. It will also get you out of almost anything that you have inadvertently gotten yourself into, which makes it a very valuable key.

When you want to begin a new program press and hold the Enable key while you push the Prog/Step key. In the window that opens up scroll down to copy and then press Enter. Find the template that you made for that station and press Enter again. Now you will have to enter the program number that you want for the new program.

Now you have made a copy of the station's template program so you'll need to open up the copy and use it to build the new program. To open up the copy hold the Enable key while you push the Prog/Step key again, only this time just type in the new program number and then push Enter and the copy should open up. If you are using the program protection as I've suggested then you'll have to remove the protection before you proceed to build the new program.

Once you have some programs in place there will be times when you will need to go in and tweak them, never make anything other than minor changes to a working program. If you are going to do anything more complex than moving a couple of points then it will pay you to make a copy and change the copy instead. There is simply no sense in taking a chance in messing up a program that you have so much time and energy invested in. Once you have the copy of the program working the way it should it's easy to replace the original with the copy. All you have to do is press both the Enable and the Prog/Step keys together, only this time scroll down to 'Rename' instead and you simply rename the copy with the number of the original and the copy will take the place of the original.

All of this may seem to be like a lot of unnecessary work but if you ever have to try to rebuild a section of programming after a failed attempt to improve a program you'll understand why I happily expend the small amount of time and effort that it takes to create a copy to work with. Believe me, I Been there and Done that and quite frankly

I'm getting more than a little tired of having to kick myself in the butt, it's well worth the extra work.

One of the neatest and handiest tricks that the newer OTC Daihen robots is able to do is shifting all or part of a program to a different location in the X, Y, Z space of an individual station. With most of the robots that I've had to deal with if you needed to move a fixture over a few inches you would have to write a new program from scratch. If you happen to be lucky enough to be programming one of these new OTC Daihen robots then it is actually a quick and easy procedure to just shift the program over to a new location.

To move a program to a different grid location you need to first determine exactly how far you need it to move. This is one reason that it is good to have a uniform grid of holes on the tables of each station having equal dimensions in both the X and Y trajectories.

I prefer a grid of 1/2 X 13 standard threaded holes in a three inch by three inch pattern. If you are designing your own cell I suggest a similar pattern with all of the holes threaded. I have worked with some cells where only every other row was threaded and if you used the other half of the holes someone had to get under the table with a wrench and hold the nut while the bolt is tightened from above. That's the stupidest arrangement that I have personally had to deal with. I'm sure that the engineer that designed it had what they thought was a perfectly good reason for doing this and I would've loved to have been able to inquire about it but no one has ever admitted to it being their idea. I suppose that if this seemingly senseless configuration had been my idea then there's a good chance that I probably would not want anyone to know or remember it either.

The only logical reason that I can think of for that configuration is that the thread less holes were for using lineup pins. That would be okay if the pins were needed or served a useful purpose, they don't. As long as you countersink the holes in the fixture and use flat head bolts the fixture will wind up in exactly the same place every time as the tapered heads seat themselves into the countersunk holes, which is the main objective. Using the pins will also make it harder and slower to get the fixtures on and off the robot. This is counterproductive and will have the end result of costing the company in the long run, not to mention the aggravation factor that you, as the programmer, will experience when the pins needed for one fixture get in the way of the next and will need to be removed, which will cost the company even more time and consequently more money. A robot station is much better off with all threaded holes.

When moving a fixture and the associated program to a different location you need to follow the same important rule as when altering a program, begin by making a copy of the program in question. After making the copy open it up and check to see if it is protected, as it should be, then go into the 'Service Utilities' and remove the protection so you will be able to move it. The computer treats the moving of a program the same as it does the altering of it, you can't do either as long as the program is protected.

Next you'll need to go back into the 'Service Utilities' and scroll down and choose 'Program Conversion' then scroll down and choose 'XYZ Shift' to open up the page where the actual move takes place.

You are already working with a copy so you can leave the program numbers alone. Next are the steps to be moved, it automatically has, as a default, all of the steps included, you will need to change this. You do not want to move home base or the curtain dodging moves at both the beginning and the end of the program as well as the cleaning position if one has been included. When you change the step numbers be sure and use the Enter key before moving to the next one or it will just change back to the default.

I have made a habit of clicking the Enter key after most changes on the teach pendant even if it isn't required, it never hurts and there are so many times that it is required that I just do it anyway, it saves me from needing to go back and re-enter the information. The only situation where this has an unwanted effect is when the soft keyboard is open; this habit has caused me to end up with an extra character in a comment line or two.

After imputing the steps that need to be moved it is time to set the dimensions of the move itself. This is the time when having that uniform grid on the tables will truly come in handy because if you know that you moved two sets of holes and the holes are on three inch centers, then you'll know that you need to move the program exactly six inches in that direction.

You can move in two directions in a single operation or even three if you are adjusting the height as well, but most of the time I usually move in only one direction at a time. There is no special reason for this but if a programmer was to make a mistake and choose the wrong direction; negative or positive, or type in the wrong dimension, it would be a little simpler to reverse one change instead of two. As with the step numbers, don't forget to click Enter or the number will not be saved.

Always look back over the page before pushing the function key for Execute to make sure that all of the information is correct. If you did happen to forget to change the steps or not use the Enter key this will give you a chance to catch it.

Once you are sure that everything is right then choose the function for Execute. You may have error messages pop up, if you do just make note of the particular steps involved and keep clicking OK by using the Enter key until you get a message saying that the conversion is complete. Those steps, along with some others will have to be tweaked as you go through the program, which is to be expected. Moving a program using the 'XYZ Shift' is not an exact science but it will get you extremely close so that all you will have to make is very minor adjustments. Walk the robot through the program and tweak each point until satisfied. You will also need to use the same graduating test run speeds that you would use for a new program to be sure that everything is operating correctly. After you modify a copy of a program I have instructed you to rename it with the original's program number, but in the case of moving a program I suggest that you leave this as two separate programs. There is always the chance that you may need to run it in the original position again and if you keep the original program it will just be a matter of loading and running the old program number again.

Note: Don't forget to change the comment line for the grid location so you will know which program is for which grid location.

Another occasion that the 'XYZ Shift' function comes in real handy is when the customer needs changes to an existing product. As a general rule, so long as the existing fixture can be modified you shouldn't have any problem simply modifying the program as well.

Usually when engineers make changes to a print they leave one fixed point of the project stationary and then they adjust all the changes from that point. You will need to determine where that point is and reference that point for all of the changes in the program. That point will be your X=0.0, Y=0.0 and Z=0.0.

Just as when you are modifying a program or moving the entire program, the same rule about working with a copy instead of the original still applies. I know that you are already getting tired of having me repeat this over and over but this is important enough that it bears repeating. It is my intention to keep driving this point home so that eventually you will no longer need to remember it because you will be doing it automatically.

This is going to be another instance when you will want to keep both copies of the program, you never know if the customer may need some more of the original design built, so choose an appropriate name that you can easily distinguish between the two. I also like to include another comment line with the revision date as well. This comes under the heading of 'Covering Your Butt', which I am a firm believer it.

It will make your job a lot easier if you have a print from the original product alongside of the print for the new model so that you can compare them while doing the actual modifications to the program. If the prints have the dimensions in fractions instead of decimals you will need to convert them first. This is because the changes must be put into the 'XYZ Shift' page in the decimal format. After dealing with this problem for so long I usually end up with a conversion chart mounted either on the side of the robot cell or a nearby wall. If you don't have a conversion chart handy then grab a calculator and simply divide the bottom number into the top and the answer will be the decimal equivalent as shown below.

1/8 = one divided by eight equals 0.125

After you have all of the dimensions in decimal form all you have to do is find the difference between the two and you will have the correct numerical value of the change to use in the program conversion. When doing the conversion I usually make a note showing the correct axis and direction of the change, positive or negative, because it's very easy to move the correct amount in the wrong direction which means having to go back and do it over again.

Next you will have to determine which steps of the program you will need to shift, most of the time this will have to be done without the fixture on the robot. This is because the fixture has already been modified but the program has not so if you try to walk the robot through the original program the torch or arm may contact and damage something, including itself.

I try to have the fixture as close as I can to where it will be located so I can use it for a reference even if I have to move it as I go to prevent unwanted contact. All you need to do is move the robot through the program so you can determine exactly which steps need to be moved and in what direction, you already have the amount from comparing the prints. Take a pen and paper and make notes showing the beginning step, direction of the move, the distance and the ending step of each move. I know this is a lot of writing, but even if you are only changing a few points you'll find that it's better to have it written down in front of you than to try and do it all from memory. Once you have done this for all of the necessary moves then jog the robot back to a point just before the first move. Next you'll need to go into the 'Service Utilities' and scroll down and choose 'Program Conversion' then scroll down and choose 'XYZ Shift' to open up the page you need.

Alright, here's another important rule for you to remember; no matter how many changes that you have to make in the program only make one change at a time. Put all of the information needed for the first move then double check all of it before you 'Execute' the change. Once the first change is complete do not move on to the next change until you have jogged the robot through that change and made sure that it is doing exactly what you expected it to do. Go back and forth through it at least twice and then if it seems to be doing right then move on to the next change. Repeat this procedure until all of the changes have been implemented and then return the robot to the home position.

Once all of the changes are in place and have been checked out thoroughly it is time to do a rough test run. Place the modified fixture, complete with the parts, in position but do not attach it to the table, you will want the fixture to be able to move if something is wrong. Put the robot into continuous operation and then using the Check Go operation jog the robot through the entire program. If you can go completely through the program without contacting the fixture or the parts and everything appears to be working properly then it is time to attach the fixture to its new place on the table. Now you need to take the robot back out of continuous operation and begin moving through the program step by step tweaking all of the points as you go. When you are finished tweaking the program to your satisfaction you will need to treat the program just like a brand new program once again using the same graduating speed in the test runs as you would with a program that you have just written.

If you happen to have a project that has several different size parts and a fixture that one end or side remains fixed while the other end or side is adjustable, there may be a better option for you to consider; Call Programs. With a call program you can write a program for the welds that remain stable and you will always use that program for that section of the fixture. Then you will need to write a different program for each configuration of the movable end of the fixture. In this case the call program will always have the program number of the stable end but the model that the customer wants will determine what program number you put into the other part of the call program.

Function #80 is the program call, when you input the program number it will also bring up the name/description of the that appears in the first comment line, this way you will know that you have put the correct program number into the program call.

For projects as I described above where one end always remains the same and the other end changes I usually use what I call a generic call program. This is a simple program with nothing but one or two comment lines, two lines that call up function #80 and an end program line. I'll input the program number of the unchanging end in the first function line and program number for the changing end in the other.

If I have two small programs that will fit on a single station I'll use the same sort of generic call program so that I can run them both simultaneously on the same station.

There may be other times when the jobs may be exactly the same except for the location of one or two parts such as a moving cross member or tube. In these cases it may be more practical to just use the function #80 program call line inside the program for that job.

This use of the program call can also be handy for temporary changes as well because you can simply delete the one line of programming to change it back.

I have noticed, in my years of experience of programming robots, that very large programs do not run as good as the smaller programs. They seem to pick up more defects over time and need to be tweaked more often. I have never found a logical reason for this and the robot manufacturers will swear to you that this is not true but I have seen it so many times that I adopted a policy to not let a program go over three hundred steps and with the exception of a couple of programs I have stuck to my guns on this. If I need a longer program then I will break it up into separate programs and link them together using the program call function. Now maybe I'm wrong and it doesn't really happen, maybe it's just an illusion, but I've never had the problem again since I adopted this policy.

Another problem that you may run into with a large program happens when you try to record a point or a command and you get an error message saying that the program is too large. Most of the newer OTC Daihen robots can still only handle a file size of 64KB. If you run into this problem the best way that I have ever found to deal with it is to make a copy of the program, find a good separation point in the program and delete everything before that point in the copy and everything after that point in the original. Then you can finish the second half and use a generic call program to link the two programs together.

I have also invented another use for the program call function that comes under the heading of 'Necessity being the Mother of Invention'. Even with my favorite robots, OTC Daihen, you can't copy and paste a section of code from one program to another. I finally figured out that I could accomplish the same thing by making a copy of the program that has the section of code that I wanted to use in another program and then, in the copy, just delete all of it except for the code I needed and the program end. Then I just use the program call function to bring that section of code, as a separate program, into the program that it is needed in. So by doing this when you run the program and it gets to the line with the program call it will call the program with the code and when it gets to the program end in that program it will simply return to the main program and complete it. Maybe the next generation of robots will be able to copy and paste from multiple programs but for now I have at least found a way to get around this limitation.

It is also possible to move a program from one station to another; it is a bit more problematic and even less exact than moving a program from one location to another on a single station, but it can be done. If you are dealing with a small program with only a few welds then you will probably be better off just writing a new program from scratch on the other station. However, if you are working with a large, more complex program then moving the program will save you enough time and effort to make the move worthwhile.

First, check to see if the table is the same distance from the robot on all of the stations. If you find that they are, then just count yourself lucky because most are not. If you have a robot cell with three stations then there is a better than a fifty-fifty chance that the table in front of the robot is either closer or further away than are the two side stations. The two side stations with stations on opposite sides are probably at least close to being equal but don't count on it being too close, I guarantee that there will be at least some difference between them however small. On the other hand, if you happen to be on one of the cells that have two stations on the same side of the robot, then most programs will not be able to be shifted between the two. The last robot that I worked with had the table in front of the robot 12.125 inches further away from the robot than the first and third stations, that table was also side shifted .318 inches towards Y+ as well. Moving from station one to station three was a little more accurate in some ways but worse in others because although those two tables were almost the same distance from the robot and neither was side shifted, one of the tables had been mounted in the cell at a three degree angle from the other which meant no matter how close I managed to get the first point of the first weld set, I still had to tweak every weld point in the program each time I moved a program into or out of that station. Needless to say; I only did this when I was forced to by circumstances that were beyond my control.

The very first thing that you need to do is to make a copy of the program that you wish to move. I know that I am beginning to sound like a broken record and I would say that I am sorry, but I'm not. This is one of the most important lessons that you need to learn about programming, don't take a chance on messing up the original, and always work with a copy.

Once you have the copy opened you'll need to go into the service utilities and remove the protection before you can begin the move. Next go back into service utilities and scroll down to program conversion. This time you'll need to scroll down to the angle conversion instead of the (X, Y, Z) Shift. Double check that the program number listed is for that of the copy and not the original. Unlike when working with the (X, Y, Z) Shift this time you'll want to move all of the steps, so you will want to leave that section alone as well. This will screw up the home position as well as the curtain moves and you will need to go back into the program and correct them but that simply cannot be helped.

When doing a station shift you are usually moving the angle either ninety or one hundred-eighty degrees. As you are now moving the entire program you will only need to change one joint, (J1). You'll also notice that you can only move it sixty degrees at a time, this is not a problem. If you need to move ninety degrees just do it twice with any two numbers that add up to ninety such as sixty and thirty or forty-five and forty-five. Of course a move of one hundred-eighty degrees will require you to move it in three steps; sixty degrees at a time.

You may have error messages pop up telling you that certain steps are outside the parameters; just make note of the steps involved and keep hitting the Enter key until you get a message telling you that the move is complete. The error messages are likely to be referencing the home position and curtain moves that I mentioned earlier and in fact fixing them is the next thing that you will need to do.

The easiest way to correct these steps is by going back and forth between the template for that station and the program that you are moving. The robot should still be sitting in the home position so you can go ahead and overwrite the two home position lines in the program now. Just move to the first joint line in the program by depressing and holding the Enable key while pressing the down arrow until that line is highlighted. Then, keeping the Enable key depressed, press the O. WRITE/REC key. You'll get a message asking you to verify that you want to overwrite that line just hit the Enter key to agree with it.

The robot's control computer will almost always try to keep you from messing up a program and so will ask you to verify most program changing commands. If you had forgotten to hold the Enable key while pressing the O. WRITE/REC key it would throw up a message telling you that you are trying to improperly use the REC key. Anytime you want to use the green command of a double switch you have to keep the Enable key depressed to tell the computer that you are attempting to use that command or it will assume that you are trying to use the main command instead. Think of the Enable key as a shift key, of course if the entire key is green then you still have to depress and hold the Enable key because those keys have no function without the Enable key and so nothing will happen.

Don't forget the other home position at the end of the program; you always need to leave the robot at the home base for the start of the next run. Now you can move all the way through the program using the method that you got to the first home position line but if it's a long program you'll probably want to take the short cut. This is done by pressing the PROG/STEP key without depressing the Enable key; this is because that the step function is the main choice of that particular key. Scroll down to the 'LAST MOVE STEP' line and hit enter, that should be the home position line, and over write it in the same way that you did the other home position line.

Next you will need to replace any curtain dodging moves in the program. Now usually a robot will have one station without these moves, this means that if you are moving to that station all you will have to do is delete the curtain moves from the program. If this is the case then don't forget that there are curtain moves and the end as well as the beginning of the program. If, on the other hand, you are moving from a station that has curtain moves to another station with curtain moves then you will have to overwrite the existing moves one at a time. To do this switch to the program for that station's template and move the program to the first joint or home line and then using the Check Go function, jog the robot into the next position. Switch back to the program that you are moving and go to the first joint line after the home position and overwrite it. Using the step function go to the 'LAST MOVE STEP' again and then go to the joint line previous to it and overwrite that line as well. Continue going back and forth until you have converted all of the curtain moves to the proper ones for this station.

If you are moving from a station without curtain moves to one that needs them you'll still need to jump back and forth between the template and the moved program but in this case you'll just need to record the curtain moves instead of overwriting points.

The next thing that you need to change will be the comment lines making sure that the station number is in the first one so that it will show up when you pull up the list of programs. The best way to do this is in the edit mode so just press the edit key, it's all white so you won't need the Enable key for this, and the edit screen will open up. Using the up and down arrow keys move to the comment line that you need to change and then the right and left arrow keys to get to the comment itself. Now all that you need to do is press and hold the Enable key while pressing the edit key to bring up the soft keyboard.

You'll notice that this time you used the Enable key with a white key; this is the about the only time that I will give you this type of instruction. Now I'm going to give you another piece of very useful information and this is one that is not in any instruction manual, most of the instructors that work for the companies that sell the robots do not even know this. You'll notice that the soft keyboard includes numbers, but I don't know why they are even there because they are not needed. To use them you would have to use the arrow keys to move to each individual number and then press enter. That's a lot of work and it's completely unnecessary as well. All you have to do is use the number keys on the teach pendant to directly input the numbers into the soft keyboard display, this is much easier and faster than actually using the soft keyboard.

I really surprised my one of my own instructors with that piece of information. We both agreed that it was probably an unintentional occurrence but a neat and very handy one.

Once you get the comment lines updated with the new station number don't forget to come out of edit mode. I know that is an obvious step that anyone would know to do but I'll guarantee that it's one that will happen to you sooner or later, happens to all of us eventually.

If you haven't had a good chance to really check and see if the station's tables are in the same place relative to the robot or not then you are about to find out. Make sure that the robot and the program are both sitting in the home position, be sure that the robot is not set for continuous operation and that the parts are clamped in the fixture securely. Also, double check that the fixture is attached to the correct grid location on the new station. Now carefully begin jogging through the program to the first point at the beginning of the first weld. As you check forward, be prepared to release the Check Go key if the robot heads for a possible collision.

If the table is in the correct place the torch should end up within an eighth of an inch from where it should be. That's where I draw the line; that or less and I just tweak the program to correct the points, any more than that and I'll go into Service Utilities and use the 'X,Y,Z Shift' and bring the program to within that eighth of an inch tolerance.

Occasionally I've had to mount the fixture in a different grid location on the station that I'm moving to. If this happens to you, or if you already know that the table in that station is in a different location than the station you are moving from, then you should use the 'X, Y, Z Shift' to correct for this before trying to jog the robot through the program.

I should also mention that there will probably be times that the only way that you will be able to get to that first point so that you will be able to find out how much adjustment is necessary is to first remove the parts from the fixture. If you can then get to the programmed point you will be able to use the 'X, Y, Z Shift' to move the program enough so that you can replace the parts and then continue tweaking the program. I have occasionally had to remove the entire fixture to get the torch to the starting point of the first weld and then try to slide the fixture back into place or as close as possible so that I can determine how far I need to shift the program to be able to mount the fixture into its proper location.

After you get the program tweaked to your satisfaction you'll need to try it out. This is another case where you need to treat the program as if it were a new program that you have just written. This means reducing the speed and bringing it up in graduating steps with the weld function off until you have it up to a hundred percent.

With any major change such as this, NEVER take a chance on crashing the robot and never assume that you have not made a mistake or forgotten something essential, it's just not worth it. Always attempt to err on the side of caution by assuming that you may have indeed done something wrong, if it turns out that you have done everything exactly right, then you haven't lost anything. This is in fact one of the few times when it's not only safe to assume something, it's actually recommended.

Have you ever thought that sometimes when checking through a program step by step it would be nice if, just for a few steps, if the robot would be set on continuous motion? This would be real handy when taking it through the open air movements in between the welds. Most programmers don't bother because it's just too much trouble switching it back and forth.

If you happen to have one of the newer OTC Daihen robots you have an advantage over many of the other programmers. If you are going step by step through the welds or other tight moves close to or inside of a fixture and then the arm moves into the clear, all you have to do is press and hold the Enable key while depressing the Check Go key. As long as the Enable key is held the robot will run continuously through the points but when the arm gets close to the next weld or close quarters, all you have to do is release the Enable key and it will stop at the next point in the program and go back to running step by step until you depress and hold the Enable key again.

Note: This will only work when using the Check Go function; it will not work in reverse when you are using Check Back.

That is just another one of those neat, useful tricks that makes these Asian robots easier to program than many of their competitor's machines and why they remain my robot of choice.

Moving fixtures or dealing with adjustable fixtures are not the only times that you need to use the 'X,Y,Z Shift', there are other times when you can use it simply to save yourself time and effort.

Just to give you an example I recently was handed a fixture that would turn out twenty finished products after each run. They were simple parts with only one weld on each one, a circular weld all the way around the end of a round tube that was sticking through a square plate.

Now I could have programmed each one separately but the tubes were so thin and the angle of attack had to be exactly perfect all the way around the circle that it would have probably taken two to three hours to write the program.

What I did was take my time and got the first one programmed perfect. Using the check weld function I even went ahead and tuned it the weld until I couldn't get it any better. Then I went into Edit mode and isolated the section of code that encompassed that weld; from the approach point before the weld to the point where the torch pulled away after the Arc End. I then made a copy of that section of code and pasted it twenty times.

If you were to Check Go through the program at this point it would trace over the same weld joint twenty times in a row and in fact I would suggest that you do that the first time or two that you attempt this just to make sure that you have done it correctly.

The next thing that I did was measure the fixture to get the center to center measurements of the circular welds and found that the two columns were four inches apart and that the welds on each row were three inches apart.

The illustration below shows the setup of the fixture.

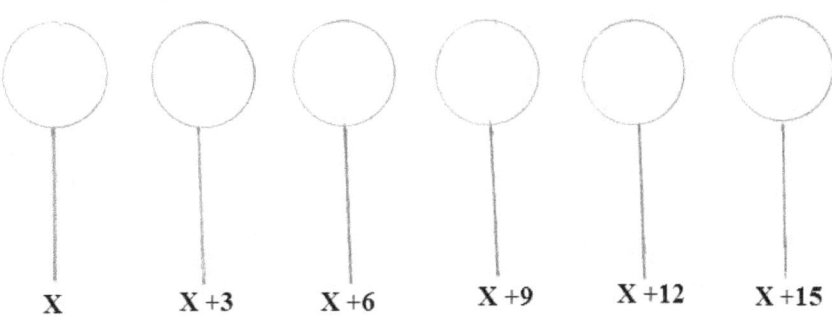

| X | X+3 | X+6 | X+9 | X+12 | X+15 |

Using the 'X, Y, Z Shift' I moved each section of code to match the fixture. I suggest that you get a pen and paper and make plenty of notes, I did and I was glad for it too. I moved the first section of code three inches, the second six inches, etc…until I reached the end of that row and then starting with that end of the second row I came back toward the first end finishing with the last section of code four inches away from the first one. After that it was just a matter of tweaking the points a bit.

Note: You can only move the programming eighteen inches at a time using the 'X,Y,Z Shift' function so if you need to move twenty-seven inches, as I did with the last shift, then you have to move that piece of code eighteen inches and then do it again for nine inches.

All together I spent about fifteen minutes programming the first weld, about thirty minutes to do the editing and another fifteen minutes to tweak the points. That meant that I spent about an hour total instead of the two to three hours it would have taken me to do the programming in the conventional way, where I would have been programming each weld one at a time. I was happy because I saved myself a lot of programming work, my boss was happy because I got the job started much faster, his boss was happy because the job went out ahead of schedule and the owner was even happier because I put even more money in his pocket than he had planned to make off of the job. Now just how many times have you ever made that many people happy with one single action?

You can also use the 'X, Y, Z Shift' in a much more localized manner. I had a project a while back that the customer wanted to switch all of the two inch square tubing over to one by two inch rectangular tubing. Now there was about twenty-five welds involved that were spread over three different fixtures and that would have taken me a long time to go in and move and overwrite all of those welds because I would've had to have physically jogged the robot through the program and moved each one individually.

By using the 'X, Y, Z Shift' I was able to do it all from within the edit mode without even moving the robot at all. All I had to do was to isolate the points involved with those two corners of the tubing and move them over two inches. It worked so well that I didn't even have to tweak any of the points after doing the test runs. Don't actually expect that to happen all of the time because it seldom will but even if I had to tweak a few of the points along the way I would've still finished the job in half the time.

Another case when the 'X, Y, Z Shift' saved me a great deal of work was a job where the customer made a change in thickness of the base plate of a product that we were welding on the robot. There were twenty-seven tubes welded to a half inch thick plate. The customer was having an issue with the product being too weak and wanted to increase the thickness of the plate to three quarters of an inch. The fixture actually would work without any adjustment at all but that meant that all of the welds needed to be a quarter of an inch higher. If I would have had to have moved each of the welds manually it would have taken me at least an hour or two but by shifting the entire program up to its new location the change only took me a few minutes. I had also created an Arc Condition file just for that program, as per my previous suggestion, so to make the change that I needed in the weld itself was simply a matter of changing one condition file instead of twenty-seven of them.

The main function of the 'X,Y,Z Shift' is as a time saver and that is what is does best even though you may sometimes need to be a little creative in finding even more uses for it.

Robot Fixture Problems

If your company is new to robotic welding you will likely have a few problems when it comes to building welding fixtures to use on the robot. The most important thing that everyone involved needs to realize is that the person that is designing and building the fixtures needs to consult the programmer throughout the entire process. Tool and Die makers that have been designing fixtures for hand welding do not understand the limitations of the robot; how far they can reach, the trouble of getting into and out of small spaces, the limits of movements and twisting, etc... If you keep the programmer in the loop you will eliminate most of the difficulties that are likely to arise if you don't.

You will always need the parts held in the fixture securely and clamps are the most common way of accomplishing this. De-sta-co makes the best fixture clamps on the market today. I know that sounds like an opinion or an advertisement but it truly is not, their clamps last much longer and work much smoother than any of their competitor's.

Even though the modern robots are now being used for much smaller production runs than many of their predecessors most of the projects given to a robot are still either jobs with high production numbers or jobs that are going to be repeat orders over long periods of time. With either of these types of cases you are going to need the kind of durable quality that you get with the De-sta-co name.

Even though you need the parts held securely in place, a clamp is sometimes not the only answer or even the best. Much of the time those clamps, no matter how bad they may be needed, simply get in the way. The MIG and TIG torches that are used on robots are much more bulky than their hand held cousins, you can't get around that. There have been so many occasions that I have had to move, bend or finally just remove a clamp so that I can get the robot in and out of a fixture that I can't even begin to count them.

The good news is that there are other options that you may not have considered or even thought about yet. One of my favorite solutions are spring clips, These really work well if you are using a lot of small parts like tabs made of sheet metal that need to be held in place. Most companies don't even have to purchase any material to make them out of. If your metal comes in the door in bundles then those bundles are likely held together with steel bands, don't throw them out. That is the perfect material to use for these spring clips; it has a very high tensile strength that will hold its shape for a long time before it will have to be replaced. The drawing below shows two of the most common shapes that I have used when forming them.

You might notice that the clips appear to be slightly over bent, this is on purpose. Over bending the clip will put a little extra pressure on the part that you need to hold in place; this can be adjusted to give you more or less pressure depending upon your need. Bolt this clip securely to the fixture and you will be able to slide the part underneath and it will be held in place while you weld it. (Hint) Always remember that you will need to be able to slide the completed part out from under the clip(s) before lifting it up.

Spring clips have another big advantage other than not having clamps in the way; it will give you a faster working fixture. Clamping and unclamping takes a great deal of time especially if the job has a lot of small parts to hold in place. I've had some fixtures where we've had upwards of twenty of these spring clips, which meant there were twenty clamps that didn't have to be clamped and released every few minutes. Another thing to look at is that these clips, made out of a material that is essentially free, are a lot lower costs than even purchasing the off-brand clamps would have been.

Another good option to use in the place of clamps is spring loaded fixtures. These fixtures require a bit more thought when designing them as well as a little more time and effort when building them but what you will get back in return for that extra effort will pay for itself many times over. Spring loaded fixtures are easier and quicker to program because you don't have to dodge or work around clamps. They also make the programs run faster because the programs generally have less points and the fixture will be much easier and quicker to load and unload. This does for the larger parts what the spring clips will do for the small tab-like parts. Below is an illustration of two different styles of devices used when building spring loaded fixtures.

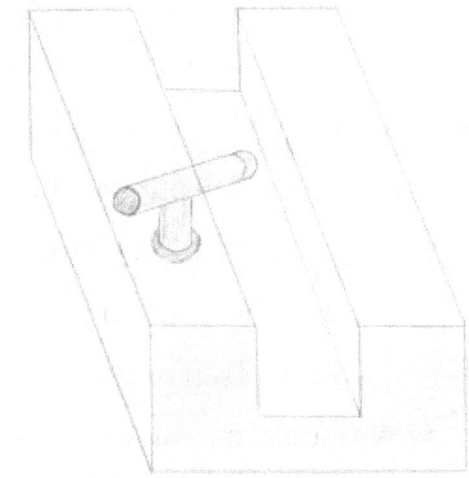

Spring Loaded Pull Clamp

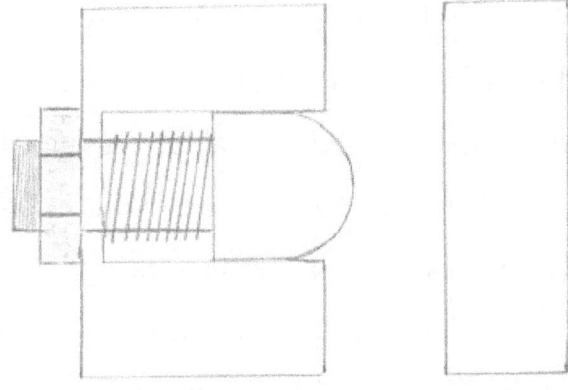

Spring Loaded Push Clamp

Another problem that I have occasionally run into is when management attempts to economize by using cheaper materials in the construction of the fixtures. In my opinion, aluminum is the material of choice for robot fixtures. The most important reason for this is because weld spatter does not stick to aluminum and if the robot is welding steel or stainless steel you can weld right down to the fixture without the chance of welding the part to the fixture.

One thing that can reduce the amount of clamps needed on a fixture or in some cases eliminate the need for them altogether is by ordered tack welding. A robot can jump from position to position so fast that so long as the tacks are put on in the correct order the drawing of the parts can be held to a minimum. This is done by always putting the tacks welds on first where the parts will be drawn against the fixture. You also don't need to tack all sides of a joint, when you get to where you would put the last tack weld that is where you start running the first full weld bead. Many times you can get by with only one tack on one side and begin welding on the opposite side. Seldom will you need over three tacks on a single joint.

Robot fixtures also need to be absolutely inflexible. If the fixture has any give to it at all you will have problems with the joints moving. A robot is going to put the weld in the exact same spot each and every time, so if the joint moves the weld is not going to be on the joint. For this reason the material that the fixture is made of needs to be thick and heavy enough so that it will be very rigid. You don't want to use bracing of any kind on a robot fixture either, even gussets. This is because the bracing will invariably limit the space that the robot can use in and around the fixture and parts. Just like with the use of clamps on a fixture the space that they take up may be needed by the robot to get to and complete the welds. That is a disadvantage that the robots have when compared to a human welder, the bulky robot arm has trouble getting into and out of the cramped spaces of the fixtures.

Another problem I have run into on numerous occasions has been the fixtures holding the part too low; this is especially true if the weld needs to wrap underneath the joint. Actually, a robot arm seems to maneuver better at a level about half way between the second and the third joints. On most average sized robot arms that would be between twelve and eighteen inches above the base mounting plate of the robot.

There are going to be occasions when drawing is going to be a factor that will need to be addressed when designing and building a fixture. Although drawing of a weld joint is, in many cases, negligible and will not adversely affect the product but in others it can and will. A certain amount of drawing cannot be avoided or prevented. All welds draw the joint to a certain degree; the heat generated by the weld as well as the width of the gap being welded will determine just how much of the drawing effect that you will experience. The type of joint that is being welded will also make a great difference in the drawing effect. It doesn't matter how many tack welds you put on the joint first or how well you have it clamped to the fixture, it will still draw some.

To demonstrate this I once welded two pieces of metal to a two inch thick table top before welding them together. Afterwards, I let it all cool completely before cutting it loose from the table and as soon as I cut the first piece loose from the table it sprang up. It had drawn even though it couldn't move at the time and then when I cut it loose the built-up pressure released all at once.

If a product is something that needs absolute precision and drawing cannot be allowed then you may have to use the drawing effect instead of trying to fight it by pre-flexing a joint in the same amount but in the opposite direction as the drawing. This can be a rather tedious trial and failure affair which will occasionally involve the scrapping a couple of parts but sometimes necessary to keep the product true and accurate.

One way to correct this problem on a fixture that has already been built is to drill and tap a hole in the fixture at the appropriate location and use a bolt or bolts, run through from the back side to pre-flex the joint prior to welding. The biggest advantage to correcting for the drawing effect in this manner is the adjustability. Generally the most difficult problem to surmount when attempting to pre-flex a joint prior to welding is accurately predicting exactly how much you need to flex the fixture, doing it this way will completely eliminate that problem.

If the company builds several products that are actually different models or different sizes of the same product there will be a temptation to build an adjustable fixture that can be used to build several different models or sizes all on a single fixture. This is actually very easy for the robot to handle; you'll simply have to write a separate program for each configuration of the fixture. The most important thing to remember when building any adjustable fixture for use with a robot is that the moveable segments must be able to be locked down securely in each position. As I said before, robot fixtures also need to be absolutely inflexible, to this end I like to have any part of the fixture that can be adjusted to be pinned and bolted tightly in place.

If the fixtures are bolted directly to the table there are going to be instances where there is simply no way for the robot arm to reach and complete all of the welds. In many cases this flaw can be reduced or eliminated through the use of positioners. The fixture is bolted to a swiveling base plate that will allow the fixture to be rotated to give the robot access to different sides of the fixture. Positioners can be electrical or mechanical in nature and can either be locked in predetermined positions or controlled by the robot programming. (You'll find out more about positioners in the next chapter.)

Even positioners can't help you complete all of the welds on some products. In cases like this another option you may use is multi-stage fixtures. In a multi-stage fixture you program all of the welds that can be made with the parts in the fixture that holds all the parts in place. After the robot runs that program you remove the parts from the first part of the fixture and put it in the second position. Usually by this time all of the separate pieces are connected so that all you need for the second stage is one clamp and a couple of stops to locate off of. This can be repeated for as many stages as necessary. I personally have only programmed four fixtures with three stages and none with more, but a lot with two.

After you write a program for each stage you'll need to write a call program to link them all together. Once you get the first stage welded and put the call program in place you move the parts that were welded in stage one to stage two and then load more component parts into stage one again so that when you push the start button it will weld stage one of the second run and stage two of the first run. Then each time the robot comes to a halt you'll take the completed products out, move the stage one parts to the second stage and load new components into the first stage again. Multi-stage fixtures are like miniature assembly lines with component parts going in and finished products coming out of every run. Many times this is the only practical way for the robot to complete all of the welds on a project. At the end of the project you will either have to cycle it down one stage at a time or if it is a reoccurring order just make sure that you cut enough extra parts to leave the partially completed parts in each stage of the fixture so that it's ready to start up the next time with completed products coming out of the fixture on every run.

Another option that I have used occasionally is to weld smaller component parts on one station of the robot to use as the component parts to build the completed product on another station. I have had jobs where I had two stations building parts that were used in a larger fixture on the third station. In these cases each cycle of the three stations produced a finished product.

One company that I worked for had two robots sitting next to each other; a smaller three station cell and a larger single station robot with a two station sliding table. We regularly built the component parts on the smaller robot to use on the larger robot that was turning out the finished products.

There is a new kind of fixture that a firm in the UK is trying to promote. It was a concept of an English engineer and they claim that it is going to be the robot fixtures of the near future. I have saw the presentation that was conducted straight from the UK via video conferencing and I'll have to admit that I was initially very impressed but when the company I was working for at the time had me actually try it out I begin to have doubts. We never did get it to work to our satisfaction on the single application that we were trying it out on. I can still see the allure of the idea and it may work well on some other applications but because of its utter failure on that first project I never got a chance to try it out on anything else.

The concept was to use three dimensional imaging to create the product digitally in their computer program. Then the program would create a cradle using only laser cut sheet metal standing upright in an interlocked grid pattern. The sheet metal would then be assembled by opposing notches that were used to lace it together similar to the way a cardboard grid is used for cushioning inside a box or carton to keep fragile items separated. The idea is to be able to assemble the fixture, use it to build a production run and then disassemble it and store it flat on a shelf until it is needed again.

If they ever get it working they will completely eliminate the need to design and build a fixture in the machine shop as well as the need for a lot of fixture storage room. The program that designed the fixture would send it directly to the laser, the laser operator would then give it to the robot operator to assemble and use. Afterwards the robot operator would then disassemble the fixture and then store it as a small stack of sheet metal. I still think that the concept is great if they can just manage to get the bugs out of it.

There were several problems that we had with it on the single attempt that I was given to try and make it work. The first one that came off of the laser went together good but was so shaky that it just wouldn't have enough stability to use as a robot fixture. The second one was much more stable, especially after we put a couple of tack welds on it to hold it steady, but we literally had to beat it together with a dead blow hammer and we destroyed it as we tried to take it apart. Even though the second fixture was stable the cradle just didn't hold the parts close enough to hold the critical tolerances that our customers require of us.

Because of these problems the company didn't buy the program and I never got to try it on another project. I think that for a lot of applications it would work just fine, just not the one that we were trying to use it on.

Positioners

Most of the robots that I have programmed could have made good use of a Positioner but almost none of them have been equipped with one. These are wonderful additions to a robot welder; if you purchase one of the good positioners that are controlled by the robot programming it is almost like having two robots working together. If your company builds a product that gets welded on more than one side or if the welds for any reason cannot be completed from one position, then there is a good chance you need a Positioner.

You probably noticed that I said that they 'Needed' a positioner, instead of saying that they could 'Use' one, which was on purpose. Without a Positioner you would need to either build a two stage fixture where you would have to write a stop into the program and then physically switch the product to another position before allowing it to continue, or have another fixture on another station where you would have to physically move the product from station to station or actually have to finish welding it by hand. Either of these alternatives is both time consuming and costly which will usually end up costing you more money than a positioner would have, especially if you take the long view. I've actually seen a positioner completely pay for itself with the first job.

Computerized positioners are programmed along with the robot; once you have programmed all the welds that can be made from the first position you bring the robot arm up out of the way and insert a stop command into the program. Then you insert the code that moves the positioner to the next position before programming the next series of welds. This is repeated until all possible welds have been completed.

There are two basic types of Positioners; the ones that predetermined lock down points (usually four), and the better Positioners which are infinitely variable and can stop at any position in the three hundred and sixty degree circle. The best ones can also be programmed to spin at any speed and can be moved while the welding is in progress as well. This will enable the robot to put a perfect, continuous weld around a circular object such as a tank while the robot remains in exactly the same position. No matter which type of Positioner you decide to get, make sure that you get one that can be used in the horizontal as well as the vertical position. These are easily recognized because of the two mounting flanges which are ninety degrees apart. They actually make some that have three flanges so that you can mount the Positioner on a forty five degree angle and others that have a swivel base and can be located at any of one hundred and eighty degrees but I have personally never used either of those types before.

There has always been some confusion between the two terms; Positioners and Turntables. Many times they are used interchangeably; I'm guilty of this myself. Some people draw the line between computer controlled and manually operated; others claim that a turntable has lockdown points while a positioner does not or motorized vs. non-motorized. Still others tell me that it is directly related to the size, but they can't seem to tell me exactly at what size that the change in terms takes place.

I usually call them all positioners but occasionally I will refer to one as a turntable so I'm afraid that I'm not going to be able to help you decide which term to use. I suspect that the term 'Turntable' was the original name and it's been slowly changing or evolving into 'Positioner' over the years but that is just my theory, it's not based on any research on my part. There are many companies that are selling these devices that use both terms together in either the name or description by calling them a 'Turntable Positioner'. In the end you'll have to make up your own mind and decide what to call them but in my opinion it doesn't really matter. What does matter is whether you can find a use for one or not, and what size and/or design that you will need.

Positioners can be any size or shape, general purpose or even specially designed just for your application. There are a great many different styles being produced that you can purchase or you can even design and build your own. Positioners can be used as a stand-alone unit or can be used as a Head / Tailstock combination; these freewheeling partners are common and useful when building large bulky products or anything that needs to be supported on two ends. The face of the head can be as simple as a plate or bar with two or more holes and/or slots to attach the parts or fixtures with. They can also be as complex as a lathe styled three jawed chucks or heads with mechanical, pneumatic or hydraulic clamps. I helped build one positioner that was over ten feet in diameter. It was motorized, controlled by the robot and had four positive stop lockdown points. It could be used to spin large projects so that they could be welded on all sides or as a four station table running four identical fixtures or four different jobs at once. Positioners like fixtures are only limited by your imagination or your company's needs.

Positioners have been used in welding long before the introduction of robots, in conjunction with fixed welding torches. Any weldment that needs circumferential welds can benefit from the addition of a positioner. Some examples are tanks, pipe flanges, spools, sprockets and hubs. With a fixed torch the positioner required that the object being welded to be circular but when combined with a robot they can be any shape at all.

When making the decision whether to purchase a positioner or not it actually comes down to some very basic questions.

How much will it cost to purchase and set up?

Will it improve the production speed?

Will it improve the quality of the product?

Can it be used on several different jobs that you build?

Will the return on the investment offset the cost?

How long will it take to pay for itself?

I've worked for quite a few companies that didn't own a positioner but I have never worked for one that didn't need one. On the other hand, I have never worked for a company that regretted purchasing one either. The problem is in finding a way to convince them of how much money a Positioner can put in their pocket or just how much money or how many customers they are currently losing by not having one. This can be tricky but I have found that I have had the most success when I could find a way to make them think it was their idea in the first place. I know you won't get credit for it that way but just think of how much work it will save you, believe me when I tell you that in the end it is well worth it.

When mating a robot to a positioner system it is usually best to use one that has been designed to be use with and controlled by the robot programming. There are several reasons for this but the most important of all is the safety factors that are designed into the systems. A traditional welding positioner that won't shut down at the first sign of resistance or trouble can be as dangerous for the robot as the user. Positioners that are built specifically for use with robotic welding systems are designed to shut down at the same exact moment that the robot does. This is very important because the positioners, like the robots are usually powered by extremely strong servo motors that don't choke down easily.

Programmer's Tool List

Welding shield

Welding gloves

Welper pliers

Metric and standard Allen wrenches

Metric and standard combination wrenches

Adjustable wrenches 8" and 12"

Telescoping magnet

Two flashlights

Sharpie markers

Silver and Red metal marking pencils

Measuring tape

Measuring scale

Calipers

Tri-square

Dead blow hammer

Ballpeen hammer

Chisel

Threading tap w/handle (to fit mounting holes on tables)

Air blow gun

Pneumatic die grinder

Regular pliers

Screw drivers (straight and Philips)

Files

Hand cleaner

Fraction to decimal conversion chart or table

This is by no means a complete list of tools that all robot programmers will need, it is just a basic list that every programmer will need, some more than others. I'm sure that as you go along you will find others that you will need just as much, specialized tools that your unique circumstances will demand. However, if you have everything on this list when you first start out I don't think that you'll find any of them that you won't find a use for within the first month.

Checklists

Using a checklist is a handy way to keep you from forgetting the little things that are easily overlooked. The more familiar a procedure is, the easier it is to accidentally omit essential steps in these routine activities. It's so easy to fall into the trap of thinking that you don't need to use a checklist for something that you know so well, don't let this be you, that is a recipe for disaster. Some of the highest skilled and most extensively trained professionals in the world today are the largest users of checklists. At the top of this list are surgeons and airline pilots, but anyone in a profession where they need to do the same tasks in the same order every time, needs a checklist.

Many people tend to resist using a checklist because they feel that it's an admission of fallibility or a shortcoming, others think that it will slow them down, none of this is further from the truth. In fact, using checklists to help with the routine aspects of your job is actually very liberating because it will spare you from the need to remember all of the mundane tasks associated with the job.

Feel free to copy any of these and use them, in fact, I strongly recommend that you do so.

Add any details that your particular robot or jobs dictate, these are just some general checklists, you may need to customize your own.

Once you begin using check lists you may see the need for others; scribble it on a scrap paper, write it on the wall or the side of the cell. Never feel bad about using a check list they are just reminders to prevent you from making a critical omissions from an important sequence.

Many times if I have a special job with special parts and/or steps in it then I'll make up a special checklist and tape it above the Start/Stop button so that I'll go over it before starting each run.

Using checklists have saved me from scraping a lot of material over the years as well as many hours of unnecessary re-work and even uncomfortable confrontations with the boss.

New Program Checklist

o   Secure the fixture to the table

o   Make a copy of the station template

o   Choose a program number that has meaning to the job

o   Create a recognizable name and place in the first remark line

o   Note the grid location in the program remarks

o   Create a nozzle and contact tip check point and cleaning position

o   Install a new contact tip

o   Make sure that the end of the wire is the proper distance beyond the end of the contact tip

o   Write the program checking back often

o   Set the program into the station

o   Make sure that the program begins and ends at the home position

o   Make sure to end the program

o   Check that all of the ARC STARTS and ENDS are in place

o   Check that all of the Weaves have an END

o   Check that all of the final approach speeds are at 75% or less

o   Change the accuracy of Lines, Circles, and tight moves to the highest level

o   Check Go through the program on continuous while watching the arm movements

o   Make sure that the weld function has been turned OFF

o   Make at least three reduced speed dry runs – 30%, 50%, and 70%

o   Make a full speed run

o   Reset the counter to zero

o   Double check that the gas has been turned on

o   Turn the weld function back on

o   Check each different type weld on the first run

o   Tweak the welds and/or movements as necessary

o   Organize the component parts to within easy reach

Job Begin Checklist

- Secure the fixture to the table, double check the grid location

- Organize the component parts to within easy reach

- Install a new contact tip

- Set the program into the station

- Reset the counter to zero

- Manually jog the robot through the program until satisfied that it is safe to run

- Double check that the gas is turned on

- Stop after the first completed weld and check before allowing the robot to complete run

- Check the first completed part thoroughly before continuing

Job End Checklist

o   Double check counter to make sure the job is complete

    o   Dispose of any leftover component parts

    o   Remove the fixture from the table

    o   Clean the fixture

    o   Store the fixture

    o   Clean the table

    o   Re-plug the threaded holes on the table

    o   Clear the counter in preparation for the next job

    o   Check the consumables, replace as needed

Check Weld Checklist
(OTC Daihen programmers only)

o   Put the robot in teach mode

o   Jog the robot to the point immediately preceding the weld to be checked

o   Put the robot in continuous mode

o   Make sure that the robot is not in Easy Teach mode

o   Change the protection level to 'Specialist'

o   Press F1 (Change Key)

o   Press F10 (Check Weld to ON)

o   Check for warning Beeper

o   Put Welding Shield on

o   Double check that the robot is set for continuous motion

o   Press and hold Check Go to check the weld

o   Adjust weld conditions as necessary

o   When finished change the protection level back to 'User'

Troubleshoot weld out of place Checklist

- o   Check the fixture for weld spatter and trash buildup
- o   Check for proper material
- o   Check the material cut size
- o   Tip is still centered in the nozzle
- o   Change the contact tip
- o   Check the wire extension length
- o   Check the angle of attack
- o   Check liners for drag
- o   Check the recorded points
- o   Release the protection to change points if needed
- o   Replace the protection before returning to Playback mode

End of Day Checklist

- o   Record daily progress

- o   Reset counters to zero

- o   Clean the fixtures

- o   Clean the robot arm and tables

- o   Robot back in home position

- o   Store hand tools

- o   Clean robot area

- o   Turn gas off

- o   Power down the robot arm

- o   Power down the controller

- o   Power down the welder

Programming Tips

When dealing with robots and their programming NEVER assume anything, if you do you are asking for trouble. Double check, triple check, keep checking as many times as necessary for you to be sure that nothing has been omitted. Even after all of that, I will assure you that there WILL be times when something will be overlooked and not caught. The very best that you will be able to do is to minimize the times that it will happen.

NEVER override the safety measures and get inside with the robot for ANY reason while a program is running. Funerals cost way too much, your family cannot afford one.

If the robot has an automatic function to back up the files and programs to a disc or separate drive then you should take advantage of it. Once, while working with an older robot that didn't have this function, the robot experienced a malfunction that caused a memory wipe and we lost all of the programs, over a hundred of them and they all had to be reprogrammed from scratch. This is not something that you want to take a chance on if there is a way to back up all of that hard work.

Always keep the contact tips and nozzles tight and clean. Good tips and nozzles equal good welds. This does not mean snugged up finger tight either, use the pliers and get them TIGHT. If the tip becomes loose the wire will not be aimed at the programmed position. If the nozzle becomes loose it may touch the contact tip and ground out or it may catch on the work and bring the robot to a halt, neither is acceptable.

Listen carefully to the sound of the welding arc any time the robot is working, this is usually the best way to get the first indication that the welds are not right. This hasn't changed since the days of stick welding when they told us to listen for the sound of bacon frying. Well the sound is different but the principle is still exactly the same; if it doesn't sound right then something is probably wrong. If the welding process sounds wrong but there is no obvious problem then a good place to start is the contact tip. They are the most common problem you'll have with the welds and can many times be causing a problem when they still look perfectly fine. If you replace the tip and it doesn't cure the problem you can always put the old one back in and you haven't lost anything.

When you can grab a few minutes between jobs, spend a little time and elbow grease and keep the robot cleaned up. I've seen so many robots that were so greasy and grimy that you couldn't even tell what color they are or even read the name tag and that really is a shame. If you clean it a little along spending only a few minutes at a time then it's never a big job that you have to set aside a block of time for. I realize that it may not improve the efficiency of the robot or even the quality of the welds but the appearance does have meaning. If nothing else it will impress customers that the manager or owner brings around when they are trying to decide who they want to build their products. If the boss does complain that you are wasting the company's valuable time just remind him that the robot cost more than his car and ask him would he really want his car to look like the robot does right now? Of course, if his car does look like that then this would not be a valid argument.

Every week or two at the most you need to take the time to blow out the vents in both the robot's control box as well as the welding power supply unit. Regular dust and dirt can cause the filters to clog up and cause overheating in both of them. However, you are in a weld shop so the dust that you are dealing with is actually a finely ground metallic powder instead. Now everyone knows that a metallic based powder does not mix well with sensitive electronics, you might as well pour water on them it probably wouldn't hurt them anymore than the dust. At least every few months or so you should physically open up the control box and remove the cover from the welding power supply and blow them out good.

Now you might think that you will have it rough trying to keep that metal dust out of the machines, let me tell you about my last programming experience. Management decided, in their infinite wisdom, to put their shiny, brand new robot in the sanding room. So along with all of the metallic powder ground off and threw in the air by their sanding discs and wheels I also had to deal with all of the noise generated by the air and electric sanders. That was the most misplaced robot that I have ever had the pleasure of programming.

Always clean the tables well between the removal and installation of the next welding fixtures. One tiny little spatter ball or even a small amount of crude build up, no matter how miniscule, will raise the fixture up and cause the Z axis of the program to appear to be in the wrong place.

Always keep a diligent check on ALL of the robot's consumables. These parts are called consumables for a real good reason; the robot does consume them, all of them. All of them are used at widely differing rates. Some like the contact tips, it will go through much faster than the others but they will all have to be replaced eventually. All of the consumables have one very important thing in common, as they began to wear out the quality of the weld always suffers. In fact, this is the most common reason for the robot producing inferior welds. I realize that all of the various costs of any project will affect the bottom line or profit margin as it were, but welds that are under par can cause even more disastrous results up to and including losing the customer. If a weld doesn't look good it's not just the appearance that is affected, but the integrity of the weld as well. A weld that actually fails is of course a worst case scenario but it is one that must be considered at all times. An occurrence of this nature can go far above losing the customer, even law suits which can be a fatal event for one of the smaller to medium sized companies.

The operator should always keep a close check on the course of a job. A robot is blind and does not realize when one of its consumables needs replacing or if anything else goes wrong such as a problem with the gas or wire. Each and every weld should be examined carefully for flaws or defects as the product is removed from the robot. If you consistently follow this tip as a regimen you will never have two bad parts in a row.

Never let yourself be rushed through the programming phase of the project; a little extra time spent while writing the program can always be cashed in later when the program is running. Anytime you rush through a program you can and will occasionally overlook small things, not necessarily things that would cause the robot to crash but things that will affect the efficiency of the program which in turn will lower the productivity of the robot and the profitability of the job. A few of the things that can do this are; not adjusting the accuracy of the robot movements, or not using the highest safe movement speeds in the non-weld moves, or even leaving unneeded steps in the program. Take your time while writing the program and look back through the program several times before you began the playback mode. Each time you go through the program only look for one thing; accuracy, unneeded steps, speed, etc... If you try to look for multiple occurrences you are more likely to miss some of them. I have also found that when you are looking for particular things such as the proper placement of arc starts and arc ends that you may find them easier if you are looking through the program backwards from the end to the beginning. I think that this is because if you are going through it forwards you are more likely to read over something without actually catching it.

Most of the newer robots have a way to override the maximum safe operating speed of the robot. I think they do this to entice the owners and operators of the robots to use this function to increase the overall efficiency and raise the profit margin of the jobs that the robot is running. Do not be tempted by this. NEVER operate a robot faster than 100 percent. Not only do they become much more dangerous, the increase in speed always equals a decrease in the accuracy which can cause an otherwise good program to crash and damage the robot. Doing this will also cause the axis joints, belts, servo motors, and other parts of the robots to wear out and fail ahead of schedule. When you look at the cost of replacing these parts or even the robot arm itself, operating a robot at speeds above this safety margin WILL cost you more in the long run than you can save in the now, DON'T DO IT.

Make a habit of clearing the counter after the completion of each job. Always starting a new job with a clean counter will enable you to always keep track of the robot's progress through the job. There is no sense in hand counting the parts when such a readily available and highly accurate counter is at your fingertips.

Always keep all of the files protected when you are not programming. You never know who will be put pushing the little green button when you are not around. This simple little instruction can keep anyone from messing something up that you don't really want to spend the time and effort fixing. I usually go in and protect all of the files everyday whether I remember removing the protection from a program or not, better safe than sorry.

If your robot stations have tables with threaded holes then you need to plug the unused holes to keep them clean and free of trash. I've found that the best plugs to use are the silicone plugs that are used in powder coat painting to keep paint out of the threads. This will keep you from needing to trace a tap through the threads to clean them out as often. If you do feel resistance as you thread a bolt through the holes don't force it. Take the time to run a tap through and clean the threads first to prevent the bolt from damaging the threads. This is yet another example where an ounce of prevention is worth a pound of cure.

If your robot cell utilizes an air system to open and close the doors, make sure that you regularly bleed any excess water from the system, especially when experiencing periods of high humidity. The air actuators used in this type of system are not hydraulic and moisture will prevent them from working properly or at all. If you wait for the symptoms to appear; the door moving slower and slower or the robot not beginning to move as soon as the door closes, then it will probably be too late for the bleeding to help and you will likely need to begin disassembling the cylinders and valves to get it working like it should.

Always try to use the natural inclination of the welding torch when writing the program. The robot welder will always produce better welds if it is welding in its natural direction. If you force it to go against the grain, so to speak, it will be like a right handed person tying to write or hammer with the left hand. The end result is never going to be better and will seldom be as good.

Never create too many points in a program before checking back to be sure that the robot can navigate safely and smoothly through the points. One very important reason to check back regularly and often is to make sure that you didn't forget to record one of the points, a common occurrence. If you did forget to record a point it is sometimes fun trying to get back to the point you programmed before it. You'll know exactly what you did as soon as the robot heads in a direction that you didn't expect it to when you checked back. To get back to the last point that you did record, stop the robot before it hits anything, then use the XYZ movements to jog the robot out of the way and try to check back again. Keep doing this until you get the robot back to the last recorded point and go from there. It takes mere seconds to check back but precious minutes to correct each problem, so please, check back often.

NEVER preform a major alteration of an existing program, always make a copy of the program and work with it instead. If you do manage to get the copy working properly you can replace the original but if you are not able to get the copy doing what you want it to do then you simply delete the copy, nothing lost. With the majority of robots you just simply save the copy with the original's program number and it will replace it. If you alter an existing program and you fail to make the change that you were aiming for, then you will have the problem of getting the program back to its original configuration if possible, not an easy task. Never take a chance of fouling up an existing, and working, program needlessly, always work with copies. Many times I will make a copy to work with for even minor changes, better safe than sorry.

There are going to be times that you will need to insert a weave to get a weld to flow in right or burn in correctly but my advice is to leave this as a last resort. Anytime that I can manage to make a weld work without using a weave, I will. Weaves always slow down a weld, and with it, production. Weaves also add a tremendous amount of wear and tear on the robot arm. This will reduce the lifespan of the robot by a respectable margin. So taking all of this into consideration, use a weave when you must, but only then, never use it only to put a design into the weld bead, it's simply not worth it.

NEVER change an existing arc start or weave condition to correct the problem with a weld unless you created is specifically for that program. This is because other programs are likely utilizing it and if you get it right for this particular program then it will then be wrong for all of those others. To prevent this from happening always either change to another existing condition or create a new one. If you are creating a program for a project that has quite a few welds it is usually best if you create a new arc start and weave condition. This way if you need to change all of the welds utilizing them as you tune in the welds then you will only need to make one change instead of going to each of the welds and changing them one at a time. On projects that have only a few welds this is not necessary but I have wrote many programs that have dozens of welds and a few that had upwards of a hundred welds, this tip can save you a huge amount of work on these larger projects.

If you are making changes to arc starts or weave conditions to improve the quality of a weld then only make one change at a time. This is a very important rule to remember because with multiple changes you can never be sure which one worked or didn't. Stick to only one change at a time, if it improves the weld, leave it in. If, on the other hand, it does not, then change it back and try something else.

Never leave the teach pendant laying on the work table, barrel, bin or anywhere that it may get accidentally knocked or dragged off onto the floor. They are easily broken and very expensive. I like to keep a hook at each station to hang it on when I'm not using it. One time that you should be extremely careful with the pendant that is easily over looked is when you are using the check weld function that many robots have. The screen on the teach pendants are made of a thin plastic and although most are scratch resistant, they do not resist hot weld spatter very well. I saw one a few years ago that had so much spatter stuck to it that it felt like sandpaper and because most of the spatter was stuck to the bottom of the screen it also interfered with the reading of the error messages which always appear on the lower half of the screen.

You shouldn't ever do any programming with a used contact tip in the torch, they can still look brand new and may have not been in the robot but a couple of cycles and still be messed up enough to cause you to have to go back through and tweak all of the welds after you put a new tip in. This especially true if you are welding at the upper end of the heat range because the heat can cause the soft copper of the tip to warp which in turn causes the wire to curve slightly as it comes out. You are much better off changing the tip just to be safe, you can always put the old one back in once you start begin the run so you haven't lost anything.

Some customers will want the weld wrapped completely around a joint, this request is usually made for cosmetic reasons but it adds a tremendous amount of strength to the joint as well. On some of the thinner metals this can be done using only lines but it will work better if you begin and/or end with circle commands. If the customer doesn't express an opinion I will usually decide to wrap them because of the added strength. Most welds that actually fail do not break so much as tear. By wrapping the ends and corners, which makes the welds continuous, you eliminate the end which is where a tear usually begins. Using the timer to create a pause at either the beginning or end of a weld bead can also be beneficial for the same reason.

I have found that a good practice to get into when programming is to mark your tack and stitch weld locations as well as the beginning and end points of all continuous weld beads that you are going to have to meet or join with later. One reason for getting into this practice is to prevent you from repeating a weld, this is very important when writing a large program with a lot of welds. Another reason for marking the beginning and end points of your welds is to promote consistency and regularity in the length of the weld beads. If a print calls for there to be four welds, two inches long and evenly spaced, marking them out ahead of time is the only way two accomplish this accurately. To do this you will need to carefully mark the beginning and ending points keeping in mind that those points are not the actual ends of the weld beads. Those points are in fact the center of the puddle forming the weld which is the width of the bead itself. (See the illustration below.)

This means that the weld begins half of that width before the beginning point and ends half of that width after the end point. I use a simple formula that lets me program the correct length welds. Because two halves equals a whole I subtract the width of the weld from the projected length and put that much difference between the beginning and end points when I mark them out. This will allow for the leading and trailing edges of the puddle.

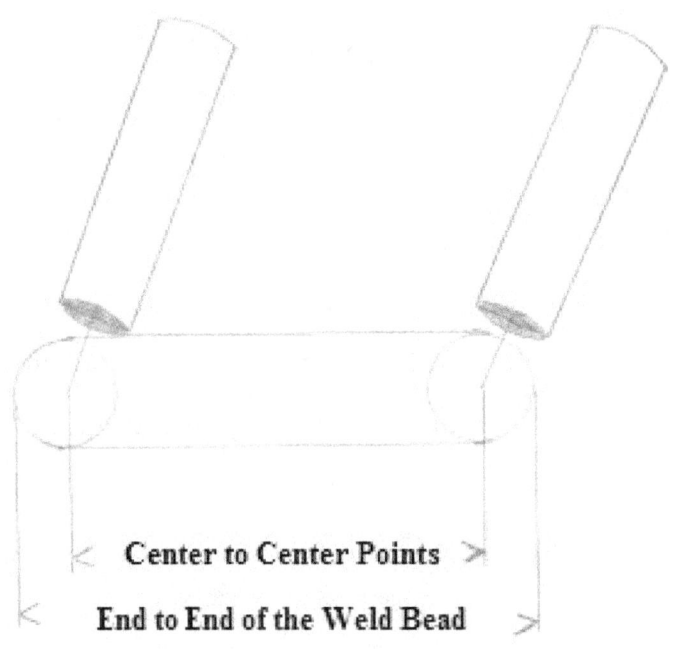

Center to Center Points

End to End of the Weld Bead

Seven rules to remember while programming

Rule number one: When you first begin to learn how to move the robot arm, DO NOT use real fixtures and parts. Use pieces of wood, cardboard boxes, toilet paper and paper towel spools, etc… get creative. This way, when, not if, when you hit them the robot will simply knock them out of the way harmlessly. Before you begin take a marker and draw around them carefully so that you'll be able to return them to the exact same spot every time. Until you can program and run the programs without moving them do not use anything that is fastened down to the table or you'll regret it. After you move all of them around several times each, you will thank me for this silly sounding rule.

Rule number two: Keep a very close watch on the speed setting especially when you have the nozzle close to the table, if it is any closer than six inches make sure that it stays on the slower speeds. This way when you slam it down on the table, and it will happen, you won't hurt anything except your pride and if no one is watching, not even that.

Rule number three: Keep a close eye on the mode of movement that the robot is set in, especially when the arm is in close quarters. If it is set in one mode but you think it is in another, then the robot is going to move in an unexpected direction and manner. This causes more robot collisions while programming than any other factor.

Rule number four: Until you are very familiar with the robot's movements and the three dimensional space that belongs to the robot DO NOT put yourself in a position where it's possible to get pinned by the robot. The dead man switch may indeed keep you from getting killed but it will not necessarily keep you from getting injured. Believe me when I tell you that the difference between a plus sign and a negative sign can and will hurt.

Rule number five: Always keep a flashlight handy. No matter how bright the lights are, where they are located, or how many there are of them, between the interference of the fixture and the component parts it will never quite get to where you need it. The great accuracy of the robots is determined by the accuracy of the points in the program. A good light pointed directly at the end of the wire while programming is essential. Repeatability is no good if the weld it is repeating is not in the right place.

Rule number six: Get in the habit of turning the gas on with the robot and welder every morning even if you are not planning on running a program that day. Once you get in the habit of switching all of them on at the same time you'll never have to worry about whether the gas is on or not before you have it weld. This small thing can keep you from throwing product in the scrap bin or grinding out bad welds to save some parts. Always remember to reverse the procedure at the end of each shift, turning all three off.

Rule number seven: Never program with a used contact tip because the wire may not be coming out completely straight. You are not going to be very happy if you have to go back through a program and move all of the points after changing to a new tip. Put a new contact tip in before beginning to write a new program no matter how little the old one has been run, it's not worth taking the chance and you can always put that one back in later on a run if it's still good.

***

Timothy Craig Everhart

Send comments or questions to:

timothyceverhart@gmail.com

www.ingramcontent.com/pod-product-compliance
Lightning Source LLC
Chambersburg PA
CBHW080239180526
45167CB00006B/2344